Advance Praise for

Mindshift

"*Significant change is possible.* With those four hopeful words, Barbara Oakley opens the door to an entirely new way of seeing and reaching our potential. Don't hesitate, it matters."
—Seth Godin,
New York Times
bestselling author of *Linchpin*

"The message of *Mindshift* is utterly convincing—you can learn, change, and grow, often far more than you can imagine. Read, learn, and enjoy!"
—Francisco J. Ayala,
professor at the University of California, Irvine,
and former president and chairman of the board
of the American Association for the Advancement of Science

"*Mindshift* is a fantastic book about how we learn and how we can use our talents (or learn new ones) in order to create a more satisfying career for ourselves. If you're stuck in a rut and don't know what to do next in life, this is a phenomenal resource to help you find your way. Dr. Oakley is a master of storytelling and of sharing ideas that can help inspire you to get out of your comfort zone and *learn*!"
—Nelson Dellis,
four-time USA Memory Champion

"Brace yourself: This book will change your entire perception of what you thought was possible. Barbara Oakley will make you realize that you can change—and change quite profoundly—by making just a few tweaks to how you learn, and she will show how these methods are becoming increasingly available to everyone. Upgrade your mind, upgrade your life, with this book."
—Scott Barry Kaufman,
scientific director of the Imagination Institute and coauthor
of *Wired to Create: Unravelling the Mysteries of the Creative Mind*

"*Mindshift* is essential reading for anyone seeking a reboot, reset, or reinvention. As Oakley trots around the globe and across disciplines, she explains the power of taking a 'pi' approach to your career, why worriers often get ahead, why negative traits can house hidden advantages, and why it's smarter to broaden your passion than follow it. Jammed with inspiring stories and practical tips, *Mindshift* is a book that can change your life."

—Daniel H. Pink,
New York Times bestselling author
of *Drive* and *A Whole New Mind*

"Following your passion is easy. Finding it is hard. This book is full of examples to help—people who have found their way around roadblocks or just plowed right through them."

—Adam Grant,
New York Times bestselling author of
Originals and *Give and Take*

"Oakley's work is remarkable for its breadth and depth . . . fascinating."

—James Taranto,
The Wall Street Journal

"Open this book to open your mind. In *Mindshift*—both a collection of inspiring stories and a field guide to creating change—Barbara Oakley shows how deep learning, deep practice, and deep transformation work and drive progress and possibilities."

—Guru Madhavan,
author of *Applied Minds: How Engineers Think*

"In an age when more and more Americans find themselves changing jobs and careers, *Mindshift* provides indispensable advice and help."

—Glenn Harlan Reynolds,
Beauchamp Brogan Distinguished Professor of Law,
The University of Tennessee

Mindshift

Break Through Obstacles to Learning and Discover Your Hidden Potential

BARBARA OAKLEY, PHD

A TarcherPerigee Book

★

An imprint of Penguin Random House LLC
375 Hudson Street
New York, New York 10014

Most TarcherPerigee books are available at special quantity discounts for bulk purchase for sales promotions, premiums, fund-raising, and educational needs. Special books or book excerpts also can be created to fit specific needs. For details, write: SpecialMarkets@penguinrandomhouse.com.

Library of Congress Cataloging-in-Publication Data
Names: Oakley, Barbara A., 1955– author.
Title: Mindshift: break through obstacles to learning and discover your hidden potential / Barbara Oakley.
Description: New York: TarcherPerigee, 2017.
Identifiers: LCCN 2016041190 (print) | LCCN 2017002428 (ebook) | ISBN 9781101982853 (paperback) | ISBN 9780399184086
Subjects: LCSH: Self-actualization (Psychology) | Continuing education. | Adult education. | BISAC: EDUCATION / Adult & Continuing Education.
Classification: LCC BF637.S4 O25 2017 (print) | LCC BF637.S4 (ebook) | DDC 158.1—dc23
LC record available at https://lccn.loc.gov/2016041190

Printed in the United States of America
1 3 5 7 9 10 8 6 4 2

Contents

Chapter 1

Transformed

Graham Keir's career was charging forward, unstoppable as a bullet train. He wasn't just following his passion—it was driving his life.

Or so he thought.

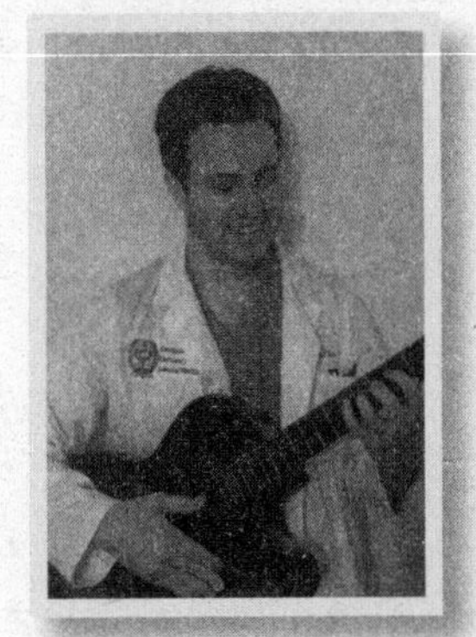

Graham Keir's career switch from the music he adored to the math and science he had loathed came as a shock even to him. Nowadays, he couldn't be happier.

Even in grade school, Graham was obsessed with music. An upbeat child, he played violin from the time he was four, then nimbly expanded his repertoire by picking up the guitar at eight. In high school, the smoky world of jazz beckoned, and he began practicing this new freeform rhythm with nearly every breath he drew.

Graham lived just outside Philadelphia, once the home of jazz greats like Billie Holiday, John Coltrane, Ethel Waters, and Dizzy Gillespie. In the evenings, he would slip away from the spacious yard of his family's old Victorian house right next to a train station and onto the clanking Southeastern Pennsylvania Transportation Authority R5 train. Disembarking onto the stained concrete in Philadelphia, he'd step into the magical world of jazz clubs and live jam sessions. It was in listening to jazz that he came alive.

Eventually, Graham would train at two of the best conservatories, the Eastman School of Music and the Juilliard School, and he would be featured in *DownBeat* magazine as Best Soloist at the college level.

This wasn't to say that Graham was a success in every area of his life. Far from it. Pretty much anything that wasn't music-related was given short shrift. Math was a frustration—he blundered through algebra and geometry and never touched calculus or statistics. His high school science record was lousy. After his final exam in chemistry class, he came home and burned all of his work in the fireplace, thrilled to have passed. The night before the SAT, while other college-bound students lay awake nervously reviewing proofs and Advanced Placement history, Graham, flaunting his academic mediocrity, went to a jazz concert.

Graham knew that he wanted to be a musician and that was that. Even the mere *thought* of math and science made him uneasy.

But then something happened. Not an accident, or a death in the family, or a sudden shift of fortune. It was something much less dramatic, which made the change all the more profound.

Mindshift

For decades, I've been fascinated by people who change career paths—a feat most often seen among the well-to-do, who have ample social safety nets. Even with plenty of support, however, a major career change can be as fraught as jumping from one high-speed train to another. I'm also interested in people who decide, for whatever reason, to learn the unexpected or the difficult—the expert in Romance languages who overcomes his deficits in math; the floundering gamer who finds a way to soar academically in competitive Singapore; the quadriplegic who shifts into graduate-level computer science and becomes an online teaching assistant. In an age when the pace of change is ever increasing, I've become convinced that dramatic career changes and attitudes of lifelong learning—both inside and outside of university settings—are a vital creative force. Yet the power of that force often goes unnoticed by society.

People who change careers or start learning something new later in life often feel like dilettantes—novices who never have a chance of catching up with their new peers. Much like wizards who think they're Muggles, they often remain unaware of their power.

Like Graham, I had a passionate contempt for math and science and did poorly in both from an early age. But unlike Graham, I didn't show any early talents or special abilities. I was a goof-off. My father was in the military, so we moved a lot, often landing at the rural margins of suburbia. Acreage on the edges, at least back then, was cheap, which meant we could have animals—big animals. Each school day ended with me dumping my books, leaping bareback onto my horse, and hitting the trail. Why would I care about academic learning or a lifelong career when I could be galloping through the afternoon sunshine?

Our household was monolithically English-speaking, and I floundered in seventh-grade Spanish class. My wise father listened to my whining and finally said: "Have you ever considered that the real problem isn't the teacher—maybe it's you?"

After we moved again, my father, surprisingly, was proven wrong. The new high school language teacher inspired me, making me wonder what it would be like to *think* in different languages. I learned that I liked studying languages, so I began to study French and German. Motivating teachers *matter.* They not only make you feel good about the material—they make you feel good about yourself.

My father urged me to earn a professional degree grounded in math and science. He wanted his children to be able to make their way in the world. But I remained convinced that math and science were outside my playbook. After all, I'd flunked my way through those subjects in elementary, middle, and high school. I instead wanted to study a language. At the time, there were no readily available college loans, so I bypassed college to enlist in the military where I could get paid to study a language. And I did learn a language—Russian.

But against all odds—and despite my early plans—I'm now a professor of engineering, firmly planted in the world of math and science. And with Terrence Sejnowski, the Francis Crick Professor at the Salk Institute, I teach the most popular online course in the world—"Learning How to Learn"—for Coursera/UC San Diego. The course is a MOOC—a massive open online course—and there were a million students from more than two hundred countries in the first year alone. By the time you read this book, we'll be accelerating well past the two million student mark. Educational outreach and impact like this is unprecedented—it is clear that people are hungry to learn, shift, and grow. My lifetime list of jobs and careers is eclectic, to say the least—waitress, cleaning lady, tutor, writer, wife, stay-at-home mother, U.S. Army officer, Russian translator on Soviet trawlers on the Bering Sea, and radio operator at the South Pole Station. I discovered, more or less by accident, that there was more power within me to learn and change than I had ever dreamed. What I learned in one career often enabled me to be creatively successful in the next phase of my life. And often, it was seemingly useless information from a previous career that became a powerful foundation for the next.

Now, as I watch millions of learners all over the world awakening to their potential to learn and change, I realize it's time for something new. We need a manifesto about the importance of *mindshifts* in producing vibrant and creative societies and in helping people to live to their full potential.

> A "mindshift" is a deep change in life that occurs thanks to learning. That's what this book is about.

A "mindshift" is a deep change in life that occurs thanks to learning. That's what this book is about. We'll see how people who *change* themselves through learning—and who bring prior seemingly obsolete or extraneous knowledge with them—have enabled our world to grow in fantastically creative and uplifting ways.

And we'll see how we all can be inspired by their examples—and by what we now know from science on learning and change—to learn and grow and achieve to our fullest potential.

Discovering Your Hidden Potential

People have unexpected twists in their career paths all the time. You sit down at your desk one morning, lean in to the day's work—and see your boss, flanked by security guards, ready to escort you from the building. Out of the blue, you've been let go, after two decades of hard-earned experience and mastery of the company's systems—systems that, like you, are being dumped.

Or . . . maybe you work for a jerk, and suddenly a joyous opportunity arises to escape the dungeon—if, that is, you're willing to learn something new and challenging.

Maybe you don't feel like you have a choice. Perhaps you are the obedient child who always followed your parents' admonitions, so you feel trapped in the luxury of your high-paying salary, nose pressed up against a window of longing for the career not chosen.

It might be that you eked your way through to a professional career in a place where good jobs were hard to come by. You wouldn't dream of taking a risk to shift careers, especially now that you've got children who will pay the price if you screw up.

Or . . . maybe your mother died the night before a critical exam, and you were one of the myriad students who failed the program in a system that seems purposefully designed to eliminate everyone possible. So you're stuck in a low-paying job.

Or . . . it could be that you graduated with your shiny new degree that you pursued like a zealot because you were determined to *follow your passion.* (That's what your friends always told you to do, after all.) And then, suddenly, you realize that your parents were right—the pay's lousy, the job's even worse, and to top it off, you have a career-change barrier in the form of a boatload of student debt to pay off.

Or . . . maybe you love your work, but you just feel there's something *more.*

Now what?

Different societal and personal situations place varying obstacles—some insurmountable—on learning new skillsets and on changing careers. But the good news is that worldwide, we're moving into a new era, in which training and perspectives that were once available only to the fortunate few are becoming available to many—with smaller personal and financial costs than ever before. This is not to say that a mindshift is easy. It's usually not. But the barriers have been lowered—in many cases and for many populations.

This availability of new ways of learning—new tools for a mindshift—is so overwhelming that the reaction has often been a collective *No, no, no, the older systems of career development and learning are fine. They're the only ones that matter! This new stuff is a flash in the pan.* But slowly—often unnoticed—the mindshift revolution grows. Such mindshifts don't just involve learning new skills or changing careers, but also changing attitudes, personal lives, and personal relationships. A mindshift can be a side activity, or a full-time occupation, or anything in between.

There's good evidence that our abilities to be successful in any given area aren't at all fixed. Stanford researcher Carol Dweck's "growth mind-set" centers around the idea that a positive attitude about our ability to change can help produce that change.[1] As adults, though, it's hard to know how this attitude plays out in real life. What kinds of changes can people *really* make in their interests, skillsets, and careers? What are the latest practical suggestions from research? And what role do new means of learning play in these processes?

In *Mindshift*, we'll follow people from all over the world who have made unusual career changes and overcome enormous learning challenges. There are profound insights from these "second chance" learners that are valuable no matter what career you might be shifting to or from or what you might be interested in learning. We'll watch people make difficult shifts from the humanities to the sciences or from high tech to the fine arts. We'll see how overcoming depression shares attributes with starting a new business; how even the world's most brilliant scientists can be forced to hit career reset buttons; and how being not so smart can turn out to be an asset when you are learning tough topics.

We'll examine people's motivation and learn the tricks they use to keep themselves on track during the often disconcerting process of major change. We'll hang out with fascinating adult learners and see how, especially in this digital age, you actually *can* teach an old dog new tricks. (Hint: video games can help.) We'll see what science has to say about the fresh perspectives that career changers and adult learners provide, and we'll learn practical ideas from neuroscience that can allow us to better understand how we ourselves can continue to grow mentally even well after we've reached maturity. We'll also meet a new group of learners—"super-MOOCers"—who use online learning to shape their lives in inspiring ways.

Mindshift is so important that countries are even devising systems to foster its growth. So we'll travel to Singapore, one of the most innovative of those countries, to learn of new strategies that can enhance our careers. Insights from that tiny Asian island will allow us to see innovative new ways around the *passion* versus *practicality* conundrum that often bedevils us.

Through this book, we'll also travel around the world to share a fun insider's perspective on learning, as seen from my perch at the top of the world's most popular course—a course devoted to learning. What does it look like to peer into a camera lens with millions of learners on the other side? You'll find plenty of practical advice about how to select the best ways to change and grow through learning, both online and in person.

But it isn't all just high tech; simple concepts like mental reframing and even taking advantage of some aspects of a "bad" attitude can do a lot to get us past the hurdles that life throws our way. Unconventional learners can give us unusual ideas to get around seemingly insurmountable obstacles.

This book tends to emphasize changes from artistic to mathematical or technological skillsets, rather than the other way around. This is because people often don't think an "artistic to analytic" change is possible. And, whether we like it or not, there are more societal tugs at present toward technology. But whatever you are interested in, you will find plenty of inspiration here—from the bus driver who overcomes depression, to the electrical engineer who converts to woodworking, to the publicly tongue-tied, mathematically gifted young woman who finds within herself a talent for public speaking.

Break Through Obstacles to Learning and Discover Your Hidden Potential—the subtitle of this book—paints a broad canvas. But that canvas is *your* canvas. As you'll see, the scope of your ability to learn and change is far broader than you might ever have imagined.

For now, though, let's return to Graham's story.

Graham's Shift

It was a simple thing, really, that kicked off Graham's career shift. One day, he was invited to play his guitar at a local pediatric cancer center. He hoped that his beloved music might boost the children's spirits. The brief visit turned into another visit, and then another. He found himself drawn to the courageous little patients, some of whose stories broke his heart. He was so moved by them that he eventually started a concert series for cancer patients.

As this unfolded, he began to discover something surprising. Playing music all day, every day, wasn't fulfilling him as a person. Somehow, the thought of caring personally for patients when they were at their most vulnerable began to feel more meaningful to him than performing for people he might never talk to or see again.

Suddenly, something clicked. Something impossibly scary: Graham decided that he would become a doctor.

He felt like a fool—there was nothing in his past to indicate that he could be successful in math and science. What made him think he could do this now?

Like many who struggle to reinvent themselves, he decided to start small in acquiring the mental tools he'd need. He signed up for a calculus class.

But he didn't just jump right into it. Several months before class began, he bought a precalculus e-book on his iPhone so he could run through the concepts while traveling to performances or commuting to school. At first, he found it disheartening. There were so many basic math concepts he had forgotten or poorly understood to begin with—*you mean there are rules for exponents?* He couldn't help but think, *Oh my God, what am I doing? I am at the top of my field in music, and I am about to start at rock bottom in medicine.*

However, he was well aware that one of his strengths—one he had built through years of practice in music—was the simple skill of persisting at difficult tasks. If he could practice for all of those hours to get into Juilliard, well, there was no reason he couldn't learn this new material. It would just take hard work and focus.

Knowledge of his strengths didn't remove his doubts—and didn't change the fact that his studies were often really, *really* difficult. Most of the people taking the calculus course were Columbia premed and engineering students who had taken it in high school and just wanted to boost their science GPA by retaking it. Graham felt like he was in a go-kart competing against seasoned race car drivers. When he mentioned to the professor that he was a musician, the professor couldn't figure out why Graham would want to take his class. But in the end, he fought his way to an A-minus. Not bad for a math-and-science loather's first college calc class!

A bit of Graham's doubt began to recede. But his own words convey the struggle he continually faced:

> I remember losing sleep before almost every exam because I thought, "If I don't get an A, I won't get into medical school. I just threw away my music career, and if this doesn't work, what will I have?"
>
> And there were reminders everywhere of what I had given up. The

night of the Super Bowl, I was studying for a double whammy of biochem and organic chemistry tests on the following Monday. I wasn't watching the Super Bowl, but I knew in the back of my mind that one of my friends was playing saxophone with Beyoncé during the halftime show. I had to stop looking at Facebook, because all I would see was fun things my friends were doing, be it tours or high-profile performances. I had made my decision and I needed to stand by it.

One of the hardest parts was well-meaning friends and family who tried to discourage me. They knew how successful I had been in music and couldn't see why I was doing what I was doing. Others suggested different careers that might not be as difficult. These friends planted seeds of doubt in my head that made it very hard to make it through the most difficult moments. I had to reaffirm why I was making the change by remembering specific moments of clarity that had steered me in this direction. At the same time, I didn't tell most of my musician friends what I was doing. I wanted to leave things ambiguous because it was important to maintain my connections in the jazz scene and be hired for performances. I was essentially pretending to be two different people.

At first, I limited my performing because I thought I needed to really buckle down and get to work. However, my second semester, I started playing a lot more. I got the exact same GPA as the semester before, but I was enjoying life so much more because I had a release from the daily routine. Performing was my socializing, income, and release all wrapped up into one activity.

The science classes were hard. When I first started, I had to get over the nausea that I naturally felt from math and science. Once I got into it, the material was fun and interesting. I actually started to enjoy the process of drawing organic chemistry figures and puzzling over math problems. I would smile or chuckle to myself when I saw a particularly clever solution in a textbook.

Still, I was not accustomed to the level of detail required in science classes. I would convince myself that the tests were unfair or that I really understood something but didn't show it on the test. I quickly realized, though, that someone in the class was surely getting those questions right that I wasn't. They must have certainly had a better understanding than I did. It wasn't the teacher's fault, but my fault.

I found that it wasn't enough to understand something once. I had to practice, just like I had on the guitar. I met with professors and asked questions in class. In high school, I never went for extra help because I was in denial that I was struggling with the material. I thought only the "slow" kids went for extra help. I realized, though, that I had to put my pride aside. The goal was to do well on the test, not look like a genius all the time.

I was fortunate enough to have read *Moonwalking with Einstein* just before taking these classes. I used several memory techniques such as loci, memory palace, to commit information to memory. I know that some people have naturally good memories for numbers and abstract ideas, but I wasn't one of them. It was important to figure out my limitations early on. Once I knew what I was working with, I could do what I needed to overcome them.

Graham decided to take the rest of the science requirements in a year and a summer. The first class was his old nemesis—chemistry. "Believe it or not," he noted, "I came out with an A. I had gotten a C+ in the easier high school version, but now that I had committed myself to learning the material, I had become a completely different student."

As he progressed, he found himself with A's in organic chemistry, biochemistry, and other tough classes that he would never have seen himself taking ten years before. Graham took the MCAT (Medical College Admission Test) one week after his last final. He is now in his third year of medical school at Georgetown University. I met him online after he took "Learning How to Learn" to further improve his medical school studies.

Graham's background in music has proven to be a boon to his medical career in both large and small ways. For example, in auscultation—diagnosing through listening to heart sounds—he found that his trained ear, which is sensitive to very fine differences in timbre and timing, allows him to pick up on those differences much faster than other people.

However, it is the general benefits of his background in music that have had the most impact. It is essential, of course, for physicians to have a solid understanding of the science and physiology of medicine. But Graham has found that it is perhaps equally important to be able to listen to patients and be empathetic. Playing in ensembles with other musicians, Graham learned to listen to the musicians around him and not just immediately interject his own musical thoughts. In a similar way, he found that giving patients space to talk and not immediately talking over them can lead to a better diagnosis as well as a better patient-physician relationship.

More than that, Graham has discovered that the characteristics needed to perform as a musician are surprisingly similar to those needed to "perform" in a patient encounter or procedure. He is coming to appreciate how his years of practice with musical improvisation spill over into his new life in medicine. He finds himself coping well with unexpected situations or emergencies in which he must use his growing expertise in new ways. The difficult switch from music to medicine has also allowed him to grow more comfortable with being pushed out of his comfort zone.

Physicians often tell medical students that in medical school, so much must be memorized that it can inadvertently set an expectation that medicine will be a cut-and-dried science. However, in practice, medicine is much more mutable and often relies on intuition and the "art" of healing. Graham already has the sense that his medical career will feel much more natural to him than to many medical students because of the time he has spent performing music.

But there is more. Graham wrote me:

> In my first year of medical school, I still faced struggles studying. One of the reasons I started taking your course on Coursera was because I

> knew something about my studying was inefficient. I was spending so many more hours than most people but not necessarily learning the material any better. Your course helped me realize that it is important to make studying an active process. I would spend hours rereading slides, but half the time I would just space out and lose focus. By using the Pomodoro technique and frequently testing myself, I am already seeing improvements.

So there you have it. It's possible to make enormous changes in your life—your "preprogrammed" passions or what you *think* you're good at don't have to dictate who you are or what you ultimately do. Along those lines, it's worth noting that people don't just want to change to go *into* medicine. Doctors have also slipped *out* of medicine into completely different fields. For example, despite his Harvard MD, Michael Crichton, the bestselling author of *Jurassic Park* and the television show *ER*, never bothered to obtain a license to practice medicine. And Sun Yat-sen, the founding father of the Republic of China, gave up his medical studies in Hawaii to become involved in the revolution.

The Pomodoro Technique

The Pomodoro technique is a deceptively simple, extremely powerful focusing technique developed by Francesco Cirillo in the 1980s. *Pomodoro* is Italian for "tomato," and the timers Cirillo recommended were often shaped like tomatoes. To do the Pomodoro, all you need to do is turn off all potentially distracting beeps or buzzers from your cell phone or computer, set a timer for twenty-five minutes, and then focus as hard as you can on what you're working on for those twenty-five minutes. When you're done (and this is equally important), allow your brain to relax for a few minutes—do a bit of web surfing, listen to a favorite song, walk about, chat with friends—anything to comfortably allow yourself to be distracted.

This technique is valuable in dealing with procrastination and keeping on track—even as it also has built-in periods of relaxation that are equally critical for learning.

You might say, "Hey, wait a minute. Graham was obviously a pretty bright guy—he just never put his effort into math and science before." *But how many of us are like that, with whatever subjects, skills, or areas of special expertise we've never seriously tried to tackle?*

How many of us, for whatever reason, go off track in our lives? And how many of us eventually find ways to turn things around through learning new skills and approaches? How many others seem to be on track career-wise, but have an itch for something new and sometimes scarily different?

Key Mindshift

The Value of the Beginner's Mind

Learning something new sometimes means stepping back to novice level. But it can be a thrilling adventure!

Many ordinary and extraordinary people have made fantastic changes in their lives by keeping themselves open to learning. You'll see how previous expertise in very different subject areas doesn't need to be a shackle to a past you are trying to escape. Instead, it can serve as a launching pad for creative career pathways in your present and future. And, as we'll discover in the chapters to come, science has much to say about why we choose the fields we do, how we can slip the bonds of biology, and how we can continue to learn effectively, even as we age.

Welcome aboard the new world of mindshift.

★ Now You Try!

Broaden Your Passion

Have you unnecessarily limited yourself by heeding common advice to *follow your passion*? Have you always done what you're naturally good at? Or have you challenged yourself with something that was really hard for you? Ask yourself: What could you do or be if you decided to instead *broaden your passion* and tried to accomplish something that demanded the most from you? What skills and knowledge could you bring with you from your past that could serve you as you really challenge yourself?

Surprisingly often, capturing your thoughts and putting them onto paper can help you discover what you really think and help you take more effective action. Grab a piece of paper, or better yet, a notebook you can use for this book, jot a header of "Broaden your passion," and then describe your answers to the above questions—whether your answers result in a couple of sentences or several pages.

We'll have plenty of brief active exercises like this throughout this book—as you'll discover, these exercises form outstanding ways to help you synthesize your thinking and learn at a very deep level. Reviewing your notebook or papers when you reach the end of this book will give you invaluable overview perspectives about yourself, your learning lifestyle, and your life's goals.

Chapter 2

Learning Isn't Just Studying

It all began to change when Claudia couldn't pee.

Life before that pee-based turning point hadn't been pleasant. In fact, it had been really tough. There she was, in her sixties, and she could rarely remember ever feeling good for more than a few weeks at a time.

The problem was depression. All of her life, she'd suffered from a major depressive disorder. Despite that, she prided herself on acting "normally" in front of others. This meant she would sometimes think, *I've got to get up . . . got to get up off this couch.* But this wasn't enough. It took saying it out loud—"I can move my legs"—to do the job.

But fighting that voice was another: *What does it matter? It's just not worth it.*

Her depression wasn't triggered by anything in particular. And although the signs were there early on, she was first diagnosed when she went off to college at age eighteen. This didn't come as a surprise. Depression spread its tentacles through her family—her father had also been severely depressed, as were some of her siblings.

It was in the genes. What could she do?

Claudia could usually get herself to her part-time job—she worked as a rush-hour bus driver for Metro King County in Seattle. She could also cook dinner and care for her family, whom she loved very much. From time to time, her doctors would prescribe a new drug. The drug might work for a while, but the result was always the same. Within a matter of months—a year at most—its effects would peter out and leave her as before: vacant.

She felt the urge to get out of the rat race—but then remembered that she was such a loser, she wasn't even *in* the rat race. What she was in was a kind of pervasive, ever-present pain. Still, she knew she couldn't kill herself. Her family meant too much to her. She would not—could not—hurt them. As her therapist, Paul, said, it would be "devastating" to them. In any case, raised with ironic guilt in the Catholic tradition, she realized that her death would just make a mess that other people would have to clean up.

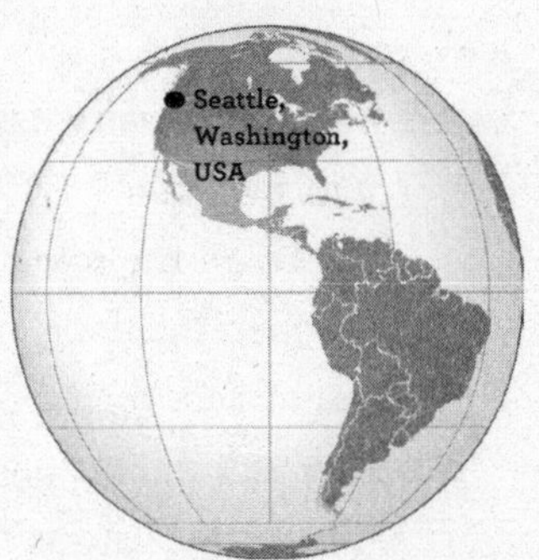

Claudia has lived in Seattle for more than fifty years—she considers herself a native of the lush, green "Emerald City."

On the job, Claudia drove either a forty-foot or a sixty-five-foot accordion-in-the-middle bus. Bus driving worked well for her because it paid decently and she could do it even when depressed. Her job was protected by the U.S. Family and Medical Leave Act of 1993; substitute drivers were built into the system. She mostly drove either morning or evening commuters. These working people formed a very different clientele from those who rode midday or late at night. The reading, dozing, functional, working-day crowds did not trigger her depression, and in any case, she avoided routes that were known for trouble and troubled people.

Still, she was living at the edge. Most people don't realize how hard it is to be a bus driver in a major metropolitan area. Buses are big, wide, and heavy. Other drivers—not to mention cyclists and pedestrians—often don't understand that it takes a lot longer for a bus to stop than a car, so they dart blithely into harm's way. There are bus-related fatalities every year in every major city. The bus drivers are almost always held responsible and usually lose their job after a serious accident.

On the morning of her accident, Claudia turned off the alarm, put on her uniform, had a quick breakfast with yesterday's coffee, and headed into the sunshine.

She signed in at work, was approved for duty, got on her assigned bus, and did a safety inspection. Drivers drove the same route, but a different bus every day. This morning, Claudia was to drive Route 308 in a forty-footer.

Once on the route, it was easy to move into the cadence of the job. Stop; open the door; wait for riders to clamber on; collect fares. . . . The bus shudders forward. Scan for passengers while watching the road. Brake, pull into a zone. Repeat.

Soon the bus was filled to capacity, with passengers standing in the aisle. Claudia steered with practiced hands onto the express lanes of the I-5 freeway. Traffic was heavy—her bus kept pace with the flow.

She was approaching the Stewart Street exit to downtown Seattle when it happened—so fast that Claudia could barely understand the sequence afterward.

Abruptly, the car in front of Claudia's bus skidded to a stop. The driver pulled toward the edge of the freeway shoulder—a narrow strip of pavement. Claudia could have swerved and barely avoided the car—except for one thing.

For no reason that Claudia was ever able to discover, the driver of the stopped vehicle opened his car door directly into Claudia's lane of traffic and began stepping out. Right in front of the bus.

Claudia glanced in her driver's-side mirror, signaled, veered left, and braked hard. It was like trying to turn and stop a twenty-ton whale balanced on a shopping cart. She found herself in the next lane—in which another car had just stopped.

She plowed into that car.

Claudia's swift reaction in slowing the momentum of her bus meant that, remarkably, none of her passengers were hurt. But after climbing down from the bus to check the car she'd hit, she realized there would be fallout to come.

Hundreds of drivers and passengers were seething in stopped cars behind Claudia's bus. After the police came, Claudia mechanically went through the procedures required after an accident. Bus drivers are supposed to drive defensively, ready to face any contingency—even bizarre contingencies like people jamming on their brakes and getting out of their car in the middle of traffic—so she was ticketed for "unsafe following distance."

It felt like a gut punch.

Claudia had been managing her depression, but she realized that this incident would knock her off the narrow ledge she'd fashioned for herself, dropping her to even darker depths. The thought was excruciating.

Meanwhile, she was hauled off by one of the bus company supervisors for drug testing. Despite the fact that she was "clean" (so clean she could have squeaked), Claudia was so stressed by the accident that she simply couldn't urinate into the little plastic cup that the lab tech from the drug-testing company had put into her hand.

After the third try, the lab tech noted in the record that Claudia "refused to produce a urine sample." Terrified, she begged for another chance. The lab tech grudgingly agreed and Claudia crept back into the stall. Desperate, she urged her body to let go.

This is it, she realized. *I am* through *with bus driving. I will deal with the ticket in traffic court. It's over.*

With those two realizations, Claudia's urine came, filling the plastic cup.

So Claudia avoided the legal imbroglio of a failed drug test. She followed through on her vow and quit her job. But there was a flip side to quitting her job: *It meant she didn't have a job.*

As predictable as the tides, severe depression rolled in. Claudia was experienced—she knew herself, and she knew exactly what lay in the months to come. The thought of so much pain, without even a job to distract her, was agonizing.

This was it. Claudia's Waterloo.

It was at this point she realized that if she wanted to escape the pain, *she* was going to have to change. Not simply vary her medicines or jobs or the little world she lived in. *She* was going to have to transform her brain, her body, her habits, and her beliefs.

Claudia was desperate, and she meant business. She told herself that she had no choice but to take her life into her own hands, since medicine and therapy weren't making life bearable. She was going to experiment with whatever she could—self-help books, teachers, coaches, cognitive neuroscience, and sheer common sense. She realized she was being melodramatic, but she was going to *learn* to get healthy unless it killed her—in one last big desperate effort at life. She was going to go through a process of discovery, experimenting on herself and keeping at it until she could see faint glimmers of light where the end of the tunnel was supposed to be.

Perky Is as Perky Does

About a month before she quit her job, Claudia was on a therapist-induced excursion to a coffee shop when she ran into an old friend who was sharing a table with another woman. The coffee shop was busy so she asked if she could join, and they readily agreed. Her tablemates had just attended their daily Jazzercise class nearby and were on an exercise high. To Claudia, exercise sounded about as fun as hammering a nail into her foot, but the women's demeanor planted a seed.

The day after the accident, instead of going to work, Claudia went to an exercise class. For the Catholic perpetrator of a bus accident, it seemed like an appropriate guilt-induced punishment.

To take part in that class, Claudia had to pay $38 for the whole month. She vowed to get her money's worth—by attending classes on every single day that she would have been working. So, as her first session unfolded, she stood in the back of the room, bopping and dipping limply along, watching as the others danced with perspiration-packed enthusiasm. Afterward, the perky instructor asked Claudia how she liked it. "I don't really move that fast," Claudia explained—to which the instructor replied, "Just try to keep up with the class," and bounced away.

But the instructor was watching.

At the next class, they shimmied. Claudia didn't know how to shimmy—after all, Catholic girls do not shimmy.

Or . . . did they?

Claudia had stepped into a new world. The class not only shimmied—they thrust out their chests and swiveled their hips while a loud, lusty male sang "Give it to me, baby." They pumped their fists in the air to the beat of "Ain't gonna let nobody get me down," and sashayed to "It's a bright, bright sunshiny day."

It wasn't long before Claudia decided she liked it.

Exercise: A Powerful (But Not All-Powerful) Tool

Claudia had tried exercise before to ward off depression, and it hadn't worked. What made her think it would work before—and why should this time be any different?

Neuroscientists used to think that you were born with all the neurons you'd ever have, and then, as you aged, neurons gradually died off. Now, of course, we realize that this is just plain wrong. New neurons are born every day, particularly in the brain's hippocampus, a vital area for learning and memory.

Kinesiology researcher Charles Hillman notes, "We've found exercise has broad benefits on cognition, particularly executive functioning, including improvements in attention, working memory and the ability to multitask."[1]

"Exercise is stronger than any medicine I could ever prescribe," Claudia's psychiatrist had told her. Indeed, exercise seems to serve as an all-purpose restart button for the brain. It does this in part by stimulating production of a protein, BDNF, which nurtures the growth of both preexisting and newly born brain cells. This effect is so powerful that it can *reverse* the decline of brain function in the elderly. Neuroscientist Carl Cotman, who did the initial breakthrough work in the area at the University of California, Irvine, has likened BDNF to a brain fertilizer that "protects neurons from injury and facilitates learning and synaptic plasticity."[2] Exercise also spurs the production of neurotransmitters—chemical messengers that transmit signals from one cell to another and one part of the brain to another. (Remember when Claudia found it so hard to get herself to move from the couch?) The simple improvement in blood flow that results from exercise may also have an effect on cognitive abilities as well as physical functioning.

As humans age, we naturally lose synapses—connecting points between neurons. It's a bit like corroding pipes that spring leaks and eventually can't bring water to where it's needed. BDNF seems to slow and reverse that "corrosive" effect. More than that, exercise seems to improve our ability to form long-term memories, although we're not sure precisely how it takes place. This, as it turns out, is a key aspect of the ability to learn. Thus, for aging brains in particular, exercise can perform the magic of a fairy godmother waving her wand.[3]

But it's important to balance the reporting here. If exercise were the only thing you needed to learn better and think more optimistically, then Olympic-level athletes should all be cheerful geniuses. Also, many people who can't exercise as a consequence of physical ailments can still learn and reason quite nicely. (Stephen Hawking seems to have done pretty well for himself.) For older individuals, walking briskly for 75 minutes a week seems to have the same positive influence on cognition as walking for 225 minutes a week.[4] (Actual fitness improves more with higher levels of exercise.) So what are we to make of this?

It seems that exercise can kick-start a cascade of neurotransmitters, along with a host of other neural changes that can shift your mind when you're trying to learn something new or think in different ways. What exercise does is set the scene to *potentiate* other changes in how your mind works. You can learn more effectively, in other words, if you've got an exercise program going on. This means, if you're serious about making a mental shift in your life, it can be invaluable to incorporate exercise into the picture.

Exercise was part of what Claudia knew she needed to do to get her out of her depressed mind-set. But she knew she needed more.

Key Mindshift
Exercise

Exercise is a powerful enhancement for any mental shift you want to make in your life. Commitment followed through with exercise has great benefits for learning and mood.

An Active Role in Changing Her Brain

Claudia had been through the drill of depressive episodes so many times. This time, she realized that if she really wanted to get herself out of the pattern, she would have to dig much more deeply than she'd ever dug before. What she'd read of how her brain worked, what she'd heard from her therapists—bits and pieces of it all resonated. She needed a mindshift to truly rewire her brain. Paradoxically, she had to be herself—but also to change in a fundamental way. To do that, she needed to make her mindshift all-important in her life.

Claudia Meadows seemed fated for a life in the shade of depression. But her active role in reshaping her thinking changed her destiny.

A dear friend of Claudia's once told her, "I've had lots of things happen that I could have gotten depressed about. I just chose to not get depressed about them. End of story." *Yeah, right,* was Claudia's reaction. *I wish.*

The notion that medicine alone would do the trick in freeing us from depression is prevalent in both doctors and the depressed—giving a pill, after all, is just so darned easy. Claudia had fallen into that trap herself—she'd once been featured in an article about the positive effects of medication on depression after pills had buoyed her for nearly a year. But soon after the article appeared, her mind shifted itself back to its well-rooted, pessimist take on life.

So Claudia's approach toward levering herself out of her bleak pit became multifaceted and determined. Just like making muscular change, making neural change would demand hard work. And lots of it.

She did some experiments, making herself go out and do things that she knew other people did for fun. *You're not so different*, she told herself. Her mind would try to do its old number on her, forecasting a glum outcome for anything she'd planned. However, she knew she couldn't always trust her mind—sometimes it told her to do stupid things. She began keeping records of her experimentation in a way that allowed for self-monitoring. Before doing something that was supposed to be fun, she would ask herself, *On a scale of 1 to 10, how much fun do I think it will be?* Afterward, she would rate herself again—and she was often surprised to see how frequently the outcome exceeded her initial expectations. In time, she began to figure out what worked for her—and she repeated what worked, whether she felt like it or not.

Claudia's Insights: Fun as a Spiritual Path

Life is full of paradoxes. For example, be authentically yourself, but change. And you don't know as much as you think you do. Read self-help books. You need all the help you can get.

- Don't always trust your mind. Sometimes it tells you to do stupid things. Find trusted advisers and run any drastic decisions by them.
- Consciously choose and acquire healthy habits. Flossing your teeth doesn't take willpower if it is a habit.

- It is much easier to imitate than initiate action. So seek advice and follow directions. Adapt it to your own circumstances. Until you can lead, follow the person in front of you. Do what they do.
- Pack your bag, purse, or gym bag the night before. You're likely to feel better about exercise the evening before than you do the morning of.
- Spend as much time outside in nature as possible. The light will do you good, and you will encounter beautiful things like plants that breathe and rocks that are proud to be rocks.
- Bring as much light to where you live as possible. Open the curtains. Use mirrors opposite windows. Use reflectors and colored glass. Be like a crow. Collect shiny objects.
- Keep going to exercise class. Eventually you will look and feel better.
- Surround yourself with lovely little things that you can afford and that make your environment beautiful. Environment counts.
- Make lists. You'll feel better if you do. And you're likely to feel even better if you do the things on the list.
- Make and hang inspirational posters, sticky notes, and pictures of people you love on your wall, and cartoons and magnets on the fridge that remind you of good times.
- You never know who is going to be your friend, so act friendly to everyone unless there is a good reason not to. Learn people's names.
- Stop complaining.

Claudia continued taking medicine, but she realized deep down that if she didn't start taking steps to rewire her thinking, her mind would slowly find its way back to the old patterns. Rewiring her brain had to be a continual daily process.

Because she was very sensitive, one of her triggers was seeing other people's suffering depicted on newscasts. So, though it was difficult, she made herself stop watching the nightly news and stop listening to the radio if a talk program or music was interrupted by news. News, after all, is primarily bad news. She began getting any necessary news and politics through a trusted friend who understood her problem.

She knew that her feelings of pain, whether she stubbed her toe or heard of another's hurt, arose only because of her own brain's perceptions. Surprisingly often, her pain came from a scary story she was telling herself about events she was viewing. Rather than being consumed by others' pain, she learned that she needed to work toward seeing the other person's problem in a rational way and asking herself if and how she could help.

Three years post–bus accident, a vibrant Claudia, age sixty-six, noted:

> A number of fortunate things came together for me after I quit my job: less stress without the bus driving; more time for sleep and taking care of myself; an opportunity for deep friendships, intellectual stimulation, and—probably most important and most difficult for me—vigorous exercise four days a week through Jazzercise, which includes upbeat music with positive lyrics.
>
> Three years after the bus accident, I'm pretty proud of myself. I could not have imagined how well I have become. I haven't gotten rich, climbed any mountains, earned any degrees, or made any momentous discoveries. But now I can get out of bed on a regular basis. I no longer feel disabling depression; I have not had a serious episode of major depression for three years and counting. I can safely say that I have learned to live my life without serious chronic, recurrent depression.
>
> I do believe that I have learned to change my way of perceiving the world to a less painful way, and that perspective has and does take continued learning and effort. I know it is not in vogue right now to emphasize the effort required to achieve what we want. Unfortunately for many of us, effort and focus is required.
>
> Living a healthy lifestyle has become my hobby and my job. I live a healthy lifestyle, not because I want to live longer, but to feel better during the time I am alive. I do not want to hurt. How do I know that

> my deliberate actions have led to increased health? I don't. As I've learned from my reading, it is not easy to rewire one's brain in the face of entrenched neural loops. How much my conscious effort is affecting my perception is not knowable to me. I choose to believe my actions do make a difference to my experience. Fun has become my spiritual path.
>
> I think that what depression has taught me is that I need to listen to myself and take care of my own needs first. Today I choose me. Then out of my abundance I can care for other people, other living beings, and then things. The learning was long and painful, but it's really quite simple. It's all about the priorities of loving.
>
> It would have been hard for the old me to believe this, but recently, a close friend told me that I am the most positive person she knows.

Indeed, I first met Claudia at a meetup in Seattle for students of "Learning How to Learn." Among the score of learners who congregated in the coffee shop, Claudia's vibrant, can-do cheerful attitude stood out. We hit it off immediately.

Claudia's Lifelong Learning

Claudia has made enormous changes in the way her mind works—changes that many people would think were impossible for someone with her biological underpinnings and her clear, lifelong patterns. Claudia notes that learning is key: "Teach yourself. Learn that getting beyond your current state is possible. Learn to change your brain and your experience of life." Exercise underpinned Claudia's ability to learn and change.

There was one big change that Claudia made, however, that we haven't really brought out. It was a vital key to her mindshift.

We'll get to that.

★ Now You Try!

Taking Active Steps

Part of Claudia's challenge was that the depression she wanted to escape made it difficult for her to want to do what she needed to escape the depression. She was in a cycle of negatively forecasting how pleasant or worthwhile events would be when she was depressed. But she moved toward health by taking active steps. These included monitoring herself and trying out new behaviors, like exercise, to keep herself on a positive self-reinforcing cycle. This allowed her to achieve and maintain a healthier mental outlook.

What mindshift are you trying to accomplish? How could you use self-monitoring in your mindshift? What thoughts are keeping you stuck? Are you trapping yourself by thinking that you are "genetically predisposed" to being unable to learn languages or math? Do you tell yourself that you are too old to make a career change? Are you inadvertently in a self-reinforcing cycle where it feels more comfortable to just continue as you are—though it leaves you dissatisfied? What positive steps could you take and what self-testing could you do to move to a new self-reinforcing cycle that begins pushing your mind to where you want it to be? What new behaviors could you immediately start to accomplish your mindshift? What do you need to do to "get off the couch and stand up"?

Jot your answers to these questions on a piece of paper, or in your notebook, under the title "Taking Active Steps."

Chapter 3

Changing Cultures

The Data Revolution

IMAGINE IT'S THE year 1704, and you're a bright, ambitious thirteen-year-old Comanche brave on the plains of what will later be called Texas. You're coming of age in a world where everyone—*everyone*—gets around on their own two feet. No planes, no cars, no horses, no nothing. Life moves in slow motion, but you don't realize it because you've never understood there could be anything different.

But suddenly, one day, you see outsized, bizarre-looking creatures gallop up on four legs—they look like oversized antelopes with no horns. Odder yet, there are humans seated on them.

What you're seeing is what you will come to call a *tuhuya*—a horse. In an instant, you realize that there are creatures on this earth that can vastly speed up your life and everything in it. Oh, the changes in your hunting! In raiding!

More than anything on earth, you want a horse.

After your first horse-snatching expedition, your ride back home feels like a bird's flight, it's so fast. Just that few extra feet of height from a horse's back makes the whole world look bigger. You practice shooting arrows from horseback, and soon, you can zip an arrow *down* into a buffalo's chest, right behind the ribcage. Your pony works with you—he seems to intuit just where he needs to be for the shot. You and your friends find yourselves retooling your people's technology—building shorter bows, which are much more maneuverable on horseback, and cobbling together saddles with stirrups, which give steadier aim.

With your dazzling new abilities, you can quickly pull down half a dozen buffalo. You can hook a leg over the withers and slip down the side of your horse while galloping past the enemy, your horse's body shielding you from arrows.

By the time you are a grown warrior, you and your friends are dazzling masters of horsemanship, in an era and culture where horses mean everything. The Comanches, in fact, took the culture of the horse to one of the highest levels in human history—their equine expertise astonished all who knew them.[1]

Eras and cultures change—change is the only consistency. We're at yet another of the many turning points in human history. The modern-day "horse" that ushers in civilization's new world is the computer.

❦

People funneling through the traditional academic degree system often don't realize how important computers can be, not to mention the mathematical thinking that underlies their operation. They don't see this, that is, until they start job hunting and understand the skills they're missing. (Both the United States and Europe are projecting big shortages in software developers.)[2]

By the time college graduates recognize they need new skills, though, it's often all too easy to believe that they can't retool. Going back to the university to get another degree is often just not possible. Few have the time or money. However, what many still don't realize is that innovative new software and computers now allow for retraining at low cost or for free.

Let's be clear. The point of this chapter *isn't* that everyone should become a computer scientist. Instead, the key idea, much like the central idea of this book, is that, whatever you *think* you are, you are actually bigger than that. You can find a way to go beyond. And you can often get started—or even complete an entire career transition—by reinventing yourself through the constantly updated world of online materials.

Whatever you *think* you are, you are actually bigger than that—you can find a way to go beyond.

By watching prototypical career changers, you can get ideas about how you can go about reframing yourself. And you, too, can discover possibilities beyond the boundaries you've unwittingly set for yourself.

Ali Naqvi and Math: "It's Complicated"

Ali Naqvi grew up in Pakistan, where he was at the top of his class through most of elementary and middle school. He reveled in English literature, history, and social studies. But there was more—Ali's father introduced him to golf at the age of seven, and he was instantly hooked. His amateur golf career soared—he won Pakistan's national amateur championship while in middle school and began representing Pakistan in international tournaments. He started to dream of playing professional golf on the PGA Tour—the main series of golf tourneys in North America.

Ali Naqvi is a business partner at Neo@Ogilvy, which is the global media agency and performance marketing network for marketing giant Ogilvy & Mather.

But there was a shadow over part of Ali's learning. Math had always been his weakest subject—and he didn't do much better in chemistry and physics. By middle school, Ali's grades in math and science slumped below average. Ali tried to get help from his teachers, but their only response was "do more practice problems" and "work harder." His parents sent him to tutors in the evenings, but Ali found himself just imitating the solutions to the problems his tutor set out; he didn't truly understand the underlying concepts.

Ali was genuinely trying as hard as he could. But one of his biggest problems was that he simply couldn't see any connection between what he was learning in math and what he saw around him in "the real world." Perhaps as a consequence, nothing sank in. He fell further and further behind his class, and his self-image as a student became compartmentalized: He was a grade A student in English, history, and social sciences, but a C-minus student in math and science.

By the time Ali got to high school, he was in serious trouble—barely passing math. It was around this time that his father was transferred and the family moved to Singapore. Here, Ali was enrolled in an inter-

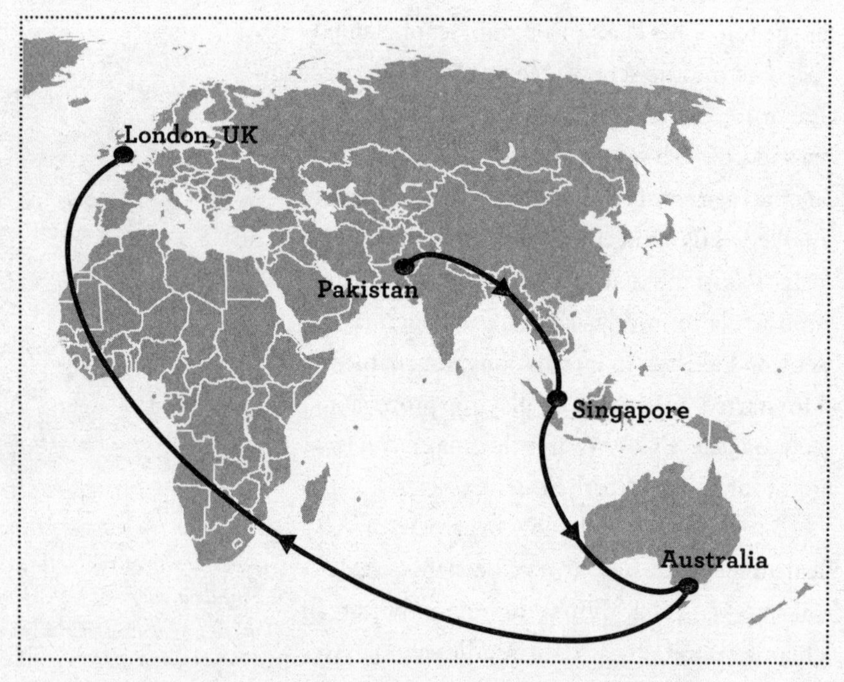

As a youngster, Ali Naqvi would never have believed how his career would unfold, and where it would take him.

national school with an American curriculum. (Pakistan follows the British system, a legacy of colonial rule.) There was an initial inching upward in his math—his new math teacher was a heavy-metal-loving ex-hippie who tricked him into learning math concepts using Metallica songs (the chorus "exit light, enter night," for example, led to balancing out two sides of an equation). But he got a new set of teachers in his sophomore year, and the terrifyingly steep-looking learning curve in precalculus and physics threw him right back into his old tailspin.

At this point, Ali basically stopped trying. He notes: "I'm not proud of it, but I accepted that I was just one of those people who was never meant to be good at math. I consoled myself by telling myself I was 'creative.' I ended up failing math and barely passing physics and chemistry. I couldn't graduate with my high school class."

It would take years for Ali to reach his educational epiphany.

Insight from Neuroscience

Becoming an expert in something new, whatever the subject, means building small chunks of knowledge using day-by-day practice and repetition. Gradually, these small chunks can then be knit together into mastery. It can seem natural to do this when learning a physical skill, say, how to play the guitar. After all, missing even one day of guitar practice can lead to fumble fingers the next day.

It may be less obvious that the same practice and repetition applies to learning in math and science. In these more cerebral "sports," you also need to practice and repeat little mental chunks. For example, after first working through a difficult homework or example problem, you can practice working the problem again from scratch without checking the solution for clues. The next day, you try this "from scratch" practice again, perhaps several times. If the problem is a tough one, you might practice it repeatedly over a number of days. You'll be surprised that what on the first day seemed completely impossible seems easy after a week's practice. "Deliberate practice" of the tougher aspects of the material allows you to develop expertise much more quickly.[3]

You can't do this with every problem, of course, but if you pick a few of the key problems to learn by heart, much like learning chords so you have them down pat, they will serve as a foundation and structure for the other material you are learning. Simply doing lots of easy problems instead of systematically stepping back to understand, practice, and repeat the toughest problems is like playing air guitar to learn how to play a real guitar.

Why is this? Insight comes in the form of this light microscopy image by biochemist Guang Yang of NYU Langone Medical Center in New York. When we learn something and then go to sleep, new synapses—vital neural connections that help us grasp and master new subjects—begin to form.[4] The triangles in the picture point toward those connections, formed overnight.

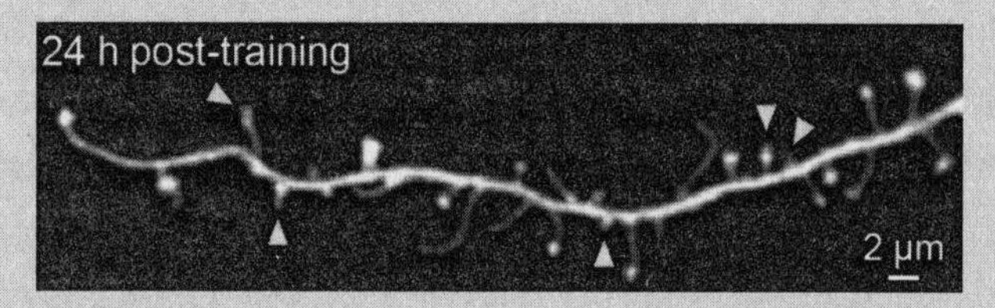

Focusing your attention on learning something, followed by sleep, is a magic combination that allows for new synaptic connections (indicated here by triangles) to form. These new synaptic connections are the physical structure that underpins your ability to learn something new.

However, only so many connections can form in a single night of sleep. This is why it's important to space learning out day by day. Additional days of practice allow for more—and stronger—neural pathways to develop.

Advanced practitioners in the STEM (science, technology, engineering, and math) disciplines realize that understanding new and sometimes difficult concepts doesn't just entail an instantaneous aha moment of understanding.[5] Moments of insight, which arise from new synaptic connections, can fade away—the connections withering—if they are not repeated soon after the original connections are formed.

Key Mindshift

Deliberate Practice of Little Chunks of Learning

Practice and repeat little chunks of learning over the course of several days. This will create the neural patterns that underlie your gradually growing expertise. The more difficult the little chunks are to learn, and the more deeply you learn them, the more rapidly your expertise will grow.

Golf: Ali's Influential Side Dream

To this day, Ali can't remember how he did it, but somehow he managed to pass the test—including the math problems—that allowed him to enter a first-year media and communication studies course in Singapore. This course served as a bridge to Monash University in Melbourne, Australia, where he ended up graduating with distinction in two and a half years.

Meanwhile, Ali didn't give up on golf. While in Australia, Ali had the chance to take lessons at the Melbourne Golf Academy from Australia's number one golf instructor—a man who coaches some of the best players in the world. Help was needed to build up the online portion of his business, and Ali suddenly found himself with a job as a web content manager.

The perks turned this into a dream job. Since Ali's office was at the driving range, he could practice his golf before and after work, as well as during lunch breaks. On weekends, he competed. Before long, he was one of the top players at his club, even competing in state championships.

But to make it to the highest levels in golf, you need to practice constantly and relentlessly. You can't hold a full-time job, as Ali had to do. Sadly, then, a career in golf didn't work out. However, Ali was to discover that his knowledge of golf would come in surprisingly handy.

A Scary Career Shift Begins

It was time for another move. This time, Ali decided to go to the United Kingdom to start a new life and a career in digital marketing—one of the few options he had from earning a media and communications degree. Two months after his move, with savings running low, Ali jumped at the opportunity to join a start-up agency as a search engine optimization (SEO) account executive, despite his lack of experience in the area.

Necessity provided a powerful push. Of all the marketing disciplines, SEO was probably the least likely for Ali to have chosen. It is one of marketing's most technical areas—demanding the math and science skills that had proved so difficult for Ali. For example, an SEO executive needs a solid understanding of servers and databases—the bricks and mortar of the Internet. It also requires an encyclopedic knowledge of SEO ranking factors, including page titles, keywords, and backlinks. Also important is knowledge of web analytics—using hard-gleaned statistical data to intuit what customers are thinking and to discover common customer "pain points" that may translate into web searches.

Most important, being an SEO executive requires knowledge of how search engine algorithms work.

Paradigm Shifts

Search engine optimization. Coding. Computers.

Change.

When it came to the revolution of the horse, we've seen how there was something special in the Comanche culture—some unusual openness to innovation and change that allowed the Comanches to seize on the benefits of the horse more quickly than other cultures. Was it a small group of mentally flexible and physically adept innovators who spread the wealth of their new horse "technology" and ideas? Possibly. Was it a pragmatism born of scrabbling at the edge of a difficult existence, where the improvement that horses offered was unusually clear? It's hard to know.

But one thing is clear—some cultures and subcultures, for better or for worse, cling more closely to past legacies. This can make it difficult for useful new ideas to scamper through the minefield of propriety and into public use. Other cultures seem more open to new ideas. But even in these more innovative cultures, great swaths of even the most intelligent people can fight off change with every fiber of their being—as evidenced among scientists by the staunch resistance to the notion of neurogenesis in adult humans, and opposition to the idea that bacteria could cause ulcers.[6] As knowledgeable academics say, moving a university is like moving a cemetery—you can't expect any help from the inhabitants.

The history of science can form a sort of relief map that allows us to see the contours of how new ideas in science, business, and culture in general can form and flow. One of the greatest analysts of the history of science was the bespectacled physicist, historian, and philosopher of science Thomas Kuhn. In examining the strands of groundbreaking scientific breakthroughs—what he called "paradigm shifts"—Kuhn noticed a pattern.[7] The most revolutionary breakthroughs, he found, were often made by one or two types of people. The first group was young people—those who had yet to be indoctrinated into the standard way of viewing matters. These individuals retained a freshness and independence of thought.

If you don't qualify as a "young person," you're probably thinking, *Well, that knocks me out, then. I'm not in my teens or twenties, so no breakthroughs for me!*

But hang on. There was a second group of people—people who were *older*, but who were as innovative as young people—people who had *switched disciplines or careers.*

It was the change in focus—the career switch—that allowed this second, older group to see with fresh eyes. Often, it also allowed them to bring their seemingly unrelated prior knowledge to the table in new ways that helped them innovate.

Old or young, you may feel like you have a childlike incompetence when you are switching disciplines. This is typical. But keep in mind that the feelings of incompetence will gradually pass—and the power you possess by virtue of your willingness to change will be invaluable.

➜ **Key Mindshift**

Switching Disciplines or Careers Brings Value

It's normal to feel inadequate when you might begin trying to understand a new subject or to broaden or change your career. Even though what you are doing is difficult, you are bringing fresh perspectives into your studies and work. This can not only be useful to your new colleagues—it can freshen your own personal outlook. *Don't discount this.*

Heading for a New Horizon

Ali's story provides an ideal snapshot of career change in midflight, as it's occurring. As you will see, the process of changing disciplines and exploring new subjects is seldom straightforward.

Ali and I had first met at a dinner in London with his one-of-a-kind colleague, advertising executive Rory Sutherland—we had admired one another's work. At that time, Ali had been full-time in digital marketing for about five years. He enjoyed what he did; however, he increasingly had the feeling that he wanted *more.* He didn't want to just give superficial advice to clients on how to create better converting websites—he wanted to have some understanding himself of what was going on under the hood. Increasingly, he came to see his day job as a bit of a tease—one that shows him on a daily basis the amazing things that those with at least a modicum of computer programming skills can accomplish.

Ali began to wonder: *Why just them—why not me?* He decided that he wasn't going to die wondering; he officially became a part of the popular "learn to code" movement.

At our first meeting, Ali told me he had been dabbling with online programming courses such as Codecademy, which many learners swear by. But like many who begin retraining themselves, he encountered his fair share of false starts. For Ali, it began to be a familiar cycle, bringing back unpleasant feelings from his battles in the past with STEM subjects: Start with enthusiasm → Make good early progress → Hit a steep learning curve when things are moving too fast → Compare himself to others who are progressing much faster → Begin to feel deflated and find excuses to procrastinate → Revisit after a while, only to find he had forgotten most of it and was back at square one.

But then he came across a book—*A Mind for Numbers*, by Barbara Oakley (yes, that's me). Ali was struck not only by the insights about learning in the book, but also by my story. I described how I'd gone from mathphobe to professor of engineering by essentially retraining my brain to grasp math and science. As Ali read about my early struggles with math, he felt I might as well have been talking about him. Ali went on to complete the MOOC "Learning How to Learn" on Coursera and gained some perspective on learning with relation to his career.

Ali's mastery of the essentials of learning, and then of coding, allowed him to feel much more comfortable with the "insides" of how computers operate. He then moved on to study web development. Perhaps subconsciously, he was building the broad intellectual tool kit he would need to underpin the more encompassing career that he truly desired.

Ali Naqvi's Practical Pointers for Effective Retraining

Here are the techniques that have been particularly useful for me:

- I have a **Pomodoro** app on my phone. This allows me to work in twenty-five-minute bursts, followed by a five-minute break. This simple technique is incredibly effective in helping me focus on *process* rather than results. The feeling of achievement after having completed my planned daily number of Pomodoros is very gratifying. I'm not perfect, but looking at my Pomodoro app stats over many months, I've consistently been waging a successful war against procrastination.
- I've found **chunking**—grasping and practicing key mental techniques until I know them like a song—to be the missing link in my search for true ownership of whatever it is that I'm learning. Giving myself a *preview* of the lesson, key concepts, and summary primes my brain for what is ahead and is like a set of support rails that frame my study sessions. Learning a new concept and then *closing my eyes and recalling* what I have just learned leaves me with no hiding place. I can't fake it anymore. If I've truly grasped it, I'll be able to recall it. If not, I go again.
- I have started **fitting in my leisure activities around my study schedule**. I feel like I am able to enjoy my favorite Netflix shows, playing the guitar, listening to music, etc., guilt-free as long as I have earned it with some focused learning beforehand. The best part about it is that I go into that leisure time knowing my brain is still working toward my goals thanks to the "magic" of the diffuse mode (neural "resting states" where you aren't thinking about anything in particular). During these "relaxed" times, I'm still learning—my brain is processing what I learned earlier.
- I've come to enjoy the practice of **applying metaphors** to concepts in whatever I am learning. I've always had quite a visual brain and a mu-

sical ear; coming up with colorful images with a fun soundtrack can even make quadratic equations fun!

- I've gotten into the habit of **thinking about the new concepts I have learned just before going to sleep**. This isn't a focused study session (or I'd never fall asleep), but rather a relaxed version of recall. I think of it as "softly opening the door to my diffuse mode." At least twice in the last couple of weeks, I've had moments of clarity when some difficult concepts hit me in the morning. I don't think this is a coincidence.
- Another technique that has worked quite well for me is the practice of **teaching myself out loud**; that is, explaining concepts to myself as if I were a complete novice. I might look like a madman talking to myself, but you quickly realize how well you do or don't understand something when you have to teach it in a concise, simple manner.

Fast-forward a year. Ali has taken a number of MOOCs related to both programming and business development, and his life has made some fascinating leaps. He has been promoted twice at his advertising agency—first to business director and now to business partner. He's fallen in love with the woman of his dreams and gotten engaged. A key theme in his life now is self-awareness. He says, "I'll soon turn thirty-two. It's clear that the best way for me to be successful is to focus on my strengths, while carefully choosing the weaknesses I want to work on. With the wedding and married life on the horizon, I'm also factoring in the added responsibilities I will have as the main breadwinner in the family."

Ali has gained a decent knowledge of web development and data analysis by virtue of his on-the-side studies. Now, at last, it has become clear that his real strength lies in knitting those recently acquired skills together with what is perhaps his greatest "value-add": relationship building. He is committed to motivating his talented team to pull together toward their shared goal. And he has long-term entrepreneurial goals in e-commerce that will meld his sporting experience and skills in digital marketing.

For a long time, Ali found it hard to forgive himself for what he saw as failures in his earlier undertakings, including his unsuccessful quest to become a professional golfer. He has since come to realize just how lucky he is to have had such rich experiences. The lessons he learned and the skills he gained are useful, not only in his current job, but in his overall career growth.

Ali Naqvi's Advice on Career Change

There will always be somebody out there who is better than you at something you want to do. You must realize that you are on your own journey, on your own path, and you are being the "best version of you" rather than a bad version of somebody else. It's normal to compare yourself to your peers; however, I think of it this way: There are a number of graphs representing different aspects of a person's life—emotional maturity, creativity, discipline, career progression, financial security, etc. These graphs don't all travel at the same trajectory for everyone. That person who kicked your butt in that golf tournament? They might kill to have your ability to play guitar. That student on the MOOC forum who seems to be able to get the programming problems so easily while you struggle? They might look at your reasoning and creative writing skills with the same level of awe as you do their ability to program. If your focus is *your truth*, you will get where you want to go when the time is right.

Focusing on the Present

One of the most valuable lessons Ali's golf coach taught him was about gaining control over his emotions and attitude. Golf can be an infuriating sport—one unlucky bounce here, one lapse in concentration there, and your chances of winning start heading south. In tournaments, when things didn't go Ali's way, he struggled to contain his frustration.

The best bit of advice his coach gave him was: "The past is the past. You can't change that. What you can control is your attitude on the next shot. The only thing in the world that matters right now is the next shot."

Applying this wisdom to online learning, Ali notes: "Online learning is an incredible privilege afforded to our generation. However, learning a complex subject such as advanced statistics or programming on your own can often be an exasperating experience. I learned this lesson during my coding program: One missed colon here and your code doesn't run. One false step in your procedure and your numbers are off. It's at moments like these where I try to follow the same standard operating procedure that I learned in my golf days—that is, acknowledge my annoyance, then take a deep breath, think about what troubleshooting steps I can take, and focus on those."

 **Now You Try!**

"Chunking" as an Important Learning Meta-Skill

Cultures are changing, and new skillsets are becoming important. Learning how to learn is an important meta-skill that can help you keep up with rapidly evolving skillsets. Ali found that mastering chunks of knowledge—how to write a brief, readable module of code, for example—was an important missing piece in his ability to gain expertise in a new area.

What is a good tiny chunk for you to practice with over several days? Give it a try and notice how it grows easier to call to mind! If you'd like, note your improvement day by day with a sentence or two on your papers or in your notebook.

Chapter 4

Your "Useless" Past Can Be an Advantage

Slipping Through Back Doors to a New Career

THROUGHOUT HISTORY, SEEMINGLY ordinary people have suddenly emerged from nowhere to seize power and shake worlds. Take Ulysses S. Grant. He was a wood-chopping lowlife, drummed from the army for drinking; yet he became one of the greatest generals of the Civil War. In more modern times, a lowly TV graphics designer in Rhode Island would emerge to become Christiane Amanpour, one of the world's leading television journalists. An adopted boy named Steve Jobs would emerge from lower-middle-class obscurity to go mano a mano against Bill Gates, who had the luxury of a world-class education from boyhood onward.

But there are more people with humble backgrounds—hundreds of millions more—who never become famous. Even so, by bringing seemingly useless knowledge from their past into their present, career changers and new learners allow society to move forward, filling needs with initially unrecognized competence.

Tanja de Bie, a project coordinator at Leiden University in the Netherlands, calls people like this "second chancers." She should know. She's one of them.

A dynamic woman with a knowing smile, a halo of hair, and an elegant Dutch lilt, Tanja exudes competence and confidence. But it wasn't always this way. People fall off the conventional university educational track for many reasons. Tanja had been a successful student, but she eventually dropped out of her history program at Leiden University to support the financial needs of her and her partner's growing family—a boy and two girls—even as her partner was also pursuing his educational goals.

Dutch administrator Tanja de Bie slowly realized that her "useless" knowledge from years of experience with her hobby provided powerful insights that helped her land her dream job.

I first spoke with Tanja over the clanking sounds of a coffee shop in Southern California—she had been flown there from the Netherlands for the online learning conference we were both attending. Tanja had the same slightly starry-eyed stare I tend to get when I'm jet-lagged and in a foreign country, but her enthusiasm was infectious.

I am struck by how similar our early life stories are—like me, Tanja had had an early passion for the humanities and a bit of a headstrong approach to life. As a backbone of financial support for her family, she worked in a secretarial capacity in various sectors—a press agency, the municipality, and health care. Though she lacked a college degree, she gradually worked her way up from secretary toward management. She was eventually drawn back to Leiden University, this time in a working capacity, by their forward-thinking philosophy of "it's what you show, not what you know." As a part of Leiden's policy de-

Tanja works at Leiden University's branch in The Hague while being a devoted resident of the nearby city of Leiden.

partment, she helped carry out various projects. But these still weren't enough for her active mind. At home in the evenings, she continued a hobby she'd developed a passion for nearly a decade before—online gaming.

Gaming

Online gaming is a very different ecosystem from that of the "real world." It involves an odd juxtaposition of analytical abilities, real-world knowledge, and people skills. Tanja found herself particularly drawn to "play by post"—a form of role-playing that focused on writing narratives in forums. Her knowledge of history gave her storytelling unusual heft—she became a vice president of one of the gaming resource communities. She also created her own online games in history and fantasy, replete with great visuals and exciting historical set pieces. The online world she inhabited was a demanding one, requiring in-depth knowledge of HTML; an ability to navigate the online legal environment; an understanding of spambots; the facility to create polls, lock topics, make global announcements; and much, much more.[1]

Tanja could turn to her hobby at home in the evening hours while also being able to drop anything at a moment's notice to be there for her young children. She found it exciting to interact through forums with people from a dizzying array of time zones around the globe. Sometimes she stayed up until the early hours of the morning, excitedly typing out stories: *"You imbecile," Le Roi muttered. His family members were giving him a headache yet again, wiping away all his hard work to encourage his English royal cousin to join his Catholic quest to the bigger glory of La France and Le Roi du Soleil. . . .*

Online gaming provided the additional excitement and creative outlet that Tanja needed in her life. Tanja is a natural storyteller, and online gaming provided an unusual creative outlet for her combined narrative and analytic talents. Tanja's excitement and joie de vivre with online gaming spilled over into her work. It was, in fact, the subject of good-natured ribbing around the coffee machine as Tanja would occasionally relate gaming mischief from the evening before.

A heartwarming aspect of the online world Tanja was involved in is the many nice people. In the real world, these are the sort of people who donate blood, serve on volunteer fire departments, or stop to assist motorists. Online, they post helpful comments on forums, help beta-test new software, and provide insightful product reviews. It's the kind of thing that reinforces a belief in humanity's innate decency.

But there's a flip side to the online world—the small percentage of people with a malevolent streak. These sinister types can have an outsized impact because of the massive online megaphone. Worse yet, the often anonymous online world has fewer of the social constraints that regulate in-person discourse. Normal people, expecting normal interactions with these more sinister sorts, are like puppies wandering up, tails wagging, in front of grizzlies.

Types known as "trolls" and "haters" enjoy creating problems in online communities. They take great glee in posting deliberately incendiary materials ("flame baiting") and in harassing and hounding others. They are also adept at creating false identities ("sock puppets") who chime in to make it seem that many are supportive of their views. Trolls can gain genuine supporters as well—often by representing themselves as misunderstood victims while praising more empathetic, kindly online users in private chats. "Haters," on the other hand, are just that—they can rant on spitefully while remaining impervious to counterarguments.

These activities can have a devastating psychological impact, not only on individual victims, but on entire online communities, which can implode into negativity, causing users to flee.

It takes a special knack—developed over time—to understand trolls, haters, and others who cause strife, and to deal effectively with them.

Tanja developed this knack through her gaming.[2]

The Challenging, Changing Needs of the Workplace

Despite the sometimes-vicious politics in academia, universities can be enjoyable places to work. Tenured academics inhabit a secure world, reigning over college students who generally understand the benefits of "playing nice" with their instructors. In face-to-face conversation, few students would ever dream of making the types of inflammatory remarks that can be tossed off anonymously online.

Also, many professors—particularly those in intense and highly technical disciplines such as medicine or engineering—are the modern equivalents of cloistered monks. These fields demand years of total dedication that can leave those in them unaware of important trends in popular culture. This means that academics—including many of the busy, world-class experts who are invited to teach massive open online courses—can suffer from curious blind spots. (We all have blind spots, and highly intelligent professors are no exception.)

One day Tanja found herself near the office's coffee machine in conversation with one of the Leiden University administrators.

The topic? Online discussion forums.

Discussion forums have long played a benign role in online education. They serve as the electronic equivalent of a coffee machine—a hub where learners can congregate and discuss the meaning of the material. Such forums have been used for decades in simple, local online classes of thirty or forty students, where there was no anonymity.

The discussion forums of MOOCs, however, are quite different. Instead of several dozen students posting on the forums, there can be thousands—and even tens of thousands—from all over the world. Tiny percentages of these students may behave badly—bullying others, uploading porn, and making threats. Others hide all sorts of vested interests, even fanaticism, that can undercut the free exchange of ideas.

Tanja was well aware of the potential for MOOCs' online forums to create incendiary problems for a university. Even a single troll or hater could change the whole tenor of the discussions. And Tanja realized that the size of MOOCs was such that multiple trolls could arise—trolls who could watch other trolls and subsequently band together to give their disruptive behavior a sense of normalcy.

As Tanja and the administrator spoke that morning by the coffeepot, the reason for the interest in discussion forums became apparent. Terrorism is a vital topic, and Leiden University was acting as a world leader in tackling the topic from an online perspective. But terrorism, especially, can serve as a lightning rod for people with sharp, fixed opinions—people unwilling to listen to any other points of view, who are willing to do whatever they can to smear those who disagree with them. So a terrorism MOOC could serve as an attractor for trolls and haters—like those Tanja had so much experience with in the world of online gaming.

Tanja couldn't help but ask—how, in the university's upcoming course on terrorism, was the university planning to handle trolls?

She was alarmed by the response: "What's a troll?"

A Gentle Aside About Gender

Tanja has an innate love of history and a natural gift for language. But she also has sharp analytical skills that reveal themselves in her love of the mechanics of games and in her participation in the world of online gaming. She's even able to design online games—a skill that goes well beyond novice levels of computer use. Although she tends to think of herself as a humanities-oriented person, it's clear that if she'd felt like it, she could have pursued a more analytical career.

No book that talks about career choice, career switching, and adult learning is complete without touching on the differences between men and women when it comes to "natural passions." Tanja de Bie's life, and her bent for the humanities despite her obvious analytic skills, illustrates some of the ways women's abilities and interests can sometimes differ from those of men.

Although, overall, boys and girls have largely the same abilities in math, an individual girl frequently finds she is better with her verbal skills than her math, while a boy frequently finds he is better with his math skills than his verbal. These inclinations grow from testosterone, which serves as a developmental drag on children's verbal abilities. Testosterone-laden boys thus can find themselves with verbal abilities that are somewhat less than those of girls of the same age.[3] (Keep in mind that this is just an average—individuals can vary quite a bit. And while boys can catch up later, by then, their self-image has already begun to solidify.)

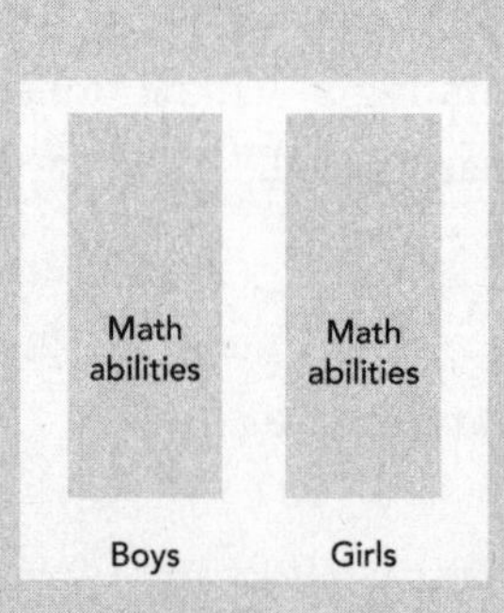

There is little developmental difference between girls' and boys' skills in mathematics as children mature.

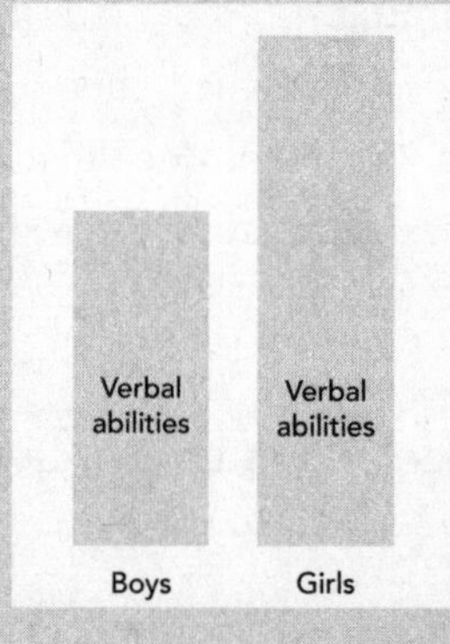

Boys in general lag behind girls in their verbal development—in toddlerdom, boys start talking later and are less talkative than girls of the same age. (This image exaggerates the average differences to make them more clear in the next figure.)

The image on the left is meant to give a sense of the developmental difference in boys' and girls' math abilities. Obviously, there is *no* real difference. But the image on the right gives a sense of the average differences in verbal abilities. Here, it is clear, boys lag behind girls.[4]

So, from toddlerhood on, girls are—on average—more verbally advanced than boys. The average boy, on the other hand, often finds that his math skills substantially outpace his verbal skills. If you put the two charts together, as below, you can see why boys frequently claim they're better at math, and girls claim to be better at verbal abilities. Both are right—even though both have the same ability, on average, for math!

We often develop passions around what we are good at. As it turns out, it often seems easier for girls to get good at subjects requiring strong verbal skills. For boys, quantitative subjects can seem easier than those involving verbal skills. Of course, testosterone can aid in muscle development, making sports seem attractive as well.[5]

Unfortunately, women's frequent big advantage—their advanced verbal skills—can inadvertently also serve as a disadvantage. Women sometimes come to believe that their passions lie solely in language-oriented areas when they also could have considerable math and science skills—on par with those of men—if only they also chose the seemingly steeper (for them) path to develop them.

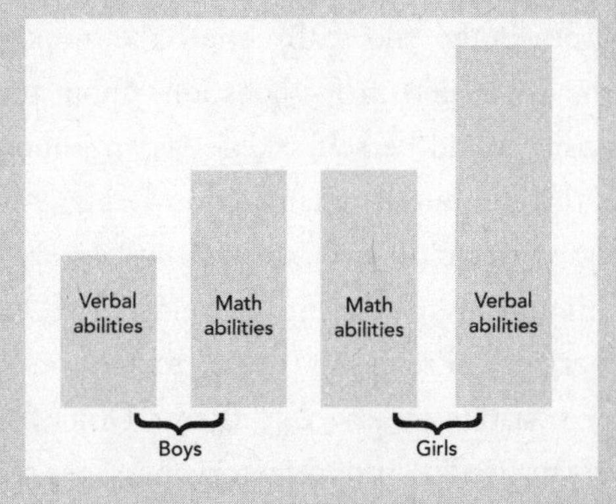

Seen together, it's clear: although boys and girls have largely the same abilities in math, an individual girl frequently finds she is better with her verbal skills than her math, while a boy frequently finds he is better with his math skills than his verbal. These inclinations grow from testosterone, which serves as a developmental drag on verbal abilities. Testosterone-laden boys are affected more than girls. (Keep in mind that this is just an average—individuals can vary quite a bit.) Although these differences fade away as children mature, early perceptions linger.

Tanja Is the Expert

"What's a troll?" Tanja couldn't believe what she was hearing. The university was about to do a MOOC on terrorism, and they were clueless about trolls?

Suddenly, the tables were turned: Tanja was no longer the lowly administrative assistant, there to support academics with years of hard-won expertise; *Tanja herself was the expert.*

That morning, Tanja began to give the administrator a feel for the dynamics of online communities, and how interactions in them are similar to and also different from those in face-to-face interactions. Tanja's concern was for Leiden University, the oldest and, in many ways, the most prestigious university in the Netherlands. She knew that without moderators overseeing the forums, the actions of just a few trolls and haters could cause these spaces to degenerate into cesspools that not only reflected poorly on the university in the press, but could also drive prospective students away.

Fortunately, Leiden's administrators knew better than to be picky about finding an academically credentialed expert. They just wanted answers—from somebody who really knew the arena. In short order, Tanja became the go-to person for questions about online forums for MOOCs. Soon, she found herself asked—as an important role in her job—to serve as the community manager on Leiden's MOOC forums. This meant bringing on volunteer mentors and training them to ensure that the tens of thousands of students in Leiden's MOOCs had a quality learning experience. Professors began to rely on her expertise as well. One of her first bits of advice to them? Don't feed the trolls. In other words, don't respond to inflammatory messages meant to provoke. And if the comments are really bad, eliminate them before they spread bad vibes through the community.

In doing this job, Tanja was influenced by the attitudes of her grandmother, a classic grand lady of the 1930s who had received an exceptional education—something quite extraordinary for that era. Her grandmother once told her, "I do not care if you get into trouble. I expect you to from time to time, as you know how to speak your mind. But let me never hear you were impolite while doing so."

Leiden's unfolding commonsense approach to dealing effectively with the self-righteous and self-important on a mission to impose their own worldview may seem obvious in retrospect, but many universities could benefit from Leiden's example. Leiden University has also been visionary in creating a new position that acknowledges the unique needs of online learning. Tanja's formal title is now project coordinator and community manager, Leiden MOOCs at Leiden University.

"Tanja created her own job," said her boss, Marja Verstelle, when I visited Leiden University at The Hague. "She is extremely dedicated. She started by spending one day a week moderating forums, but then we found she was doing more—far more. When you're looking for innovation, you're very much in need. That's the situation we found ourselves in, and Tanja was there when we needed her."

Key Mindshift

The Value of Hobbies

Hobbies often bring valuable mental flexibility and insight. If you're lucky, these insights can spill over and enhance your job. But even if they don't, your brain can be getting a workout.

Moving into a New World

Tanja's cheerful attitude, coupled with her no-nonsense abilities to take charge and make effective decisions in the new online world the university was moving into, made her a standout despite her lack of a formal college degree. As Leiden's experience with Tanja revealed, the current university credentialing system, with bachelor's, master's, and doctorate degrees, sometimes just isn't nimble enough to address the rapidly changing needs of the modern workplace with its online demands. To Leiden University's great credit, they've boldly taken the initiative in creating new jobs and filling them with the right people. This has kept Leiden from experiencing the gradual decline in MOOC enrollments that less attuned universities have experienced. Rather than declining in enrollment, in fact, Leiden University has moved to the forefront in Europe in providing quality online experiences for massive numbers of students. MOOCs are creating jobs that require experts with new skills—smart universities see this.

Tanja sometimes feels she's living in her dream. In her day job, she gets to "play" on Facebook and Twitter—getting paid to do what she likes. She also gets the perk of worldwide travel. Her opinions are valued—she is now a key figure in helping the university and major online providers facilitate events ranging from MOOCs to major international conferences.

Just as the Comanches discovered with their sudden leap into expertise in the world of horses, and as Ali Naqvi found with his shift into digital marketing, new jobs and skills are opening even as older ones fall away. But those new jobs often aren't labeled as such—and often don't even formally exist. Institutions sometimes don't even realize they need people with certain new skills, which are often so new that no one has received training in any formal program.

Tanja is a second chancer who doesn't rest in her gaming or her learning. In the past few years alone she's met with friends in London, Maryland, Pennsylvania, and California. She also games offline with dice with her children and their friends, strengthening family bonds and playing a fun and inspiring role in her children's lives.

Who knows what second—and first—chances she's building in her children's futures?

★ Now You Try!

Do You Have Special Skills or Can You Develop Them?

Over many years, Tanja de Bie developed a valuable skill in managing online communities. She wasn't afraid to use the online gaming community as a sort of career sandbox to learn, not only various aspects of programming and the mechanics of website construction, but most important, how large online communities develop and interact. Luckily, Leiden University was visionary enough to recognize that Tanja's abilities were just what they needed, whatever her academic pedigree.

Think about your own experiences. Do you have an unusual skill that's been overlooked, but that could be valuable? Are there new technical areas you could start now to gradually learn about that you've previously dismissed as something you weren't able to do? Jot your thought down in your notebook, or on a piece of paper, under the title "Special Skills."

Leaving a Job Often Leads to More Fulfillment

Surprisingly often, people's worst nightmare—having to leave a job they want to keep—ends up becoming one of the best things that's ever happened to them. This is was Kim Lachut's experience.

Kim was a "people person" who had found her dream job at her alma mater as a manager of alumni programs and services. She met amazing people, sometimes even celebrities, threw parties for them, and got paid to do it. What was not to like? Of course, the job required a lot of different skills—budgeting, locating and scheduling venues, arranging the catering, marketing, registration, lining up speakers—

and she had to have backup plans to prepare for unexpected contingencies. All this required lots of attention to detail along with great people skills. Kim thrived.

Then management changed. Kim's work environment became tense and stressful, and she began dreading getting up on weekdays. She realized a job change was in order, but job offers for event planners didn't seem to be flooding her way. What else could she do, though, especially since she'd done nothing but event planning for the past decade?

Kim Lachut was surprised to discover the IT world is a great fit for her strong people skills.

But precisely because Kim was a people person, she had lots of connections. She met the director of the full-time MBA program, who was looking for a program coordinator to handle advising, recruitment, and management of students in the program. This was right up Kim's alley, especially since she was already familiar with the university and its way of operating. With the job came the responsibility of being the IT system administrator.

However, there was a major challenge: Kim had zero experience with IT or computer software. But she was determined to head to a happy work environment, so, when the job was offered, she took it. After the first nerve-racking days settling in, she discovered something unexpected—the skills needed for IT were similar to those she'd used in event planning.

For example, there were steps to follow when she was planning an event, much like processes she followed in programming. All she needed to think of was how to handle the various unfolding possibilities. Attention to detail was critical. Kim says, "If a process isn't correct or the system isn't working properly, it hurts our students. In my job, I train our users not only on the software, but also in how this software affects the people who matter most—the students."

Kim discovered that the data-heavy IT field loves "people persons" like her, who can connect the dots between systems and the variety of individuals who use and are affected by those systems. Kim says: "I have become a self-proclaimed data geek who is able to incorporate my people skills by teaching others about how the system works, in a way everyone can understand."

Key Mindshift

The Value of Career "Catastrophes"

Those with broad experience in the working world often observe that being forced to leave a job makes people far happier with the new job than the old—no matter how impossible this might initially seem.

Chapter 5

Rewriting the Rules

Nontraditional Learning

ZACH CACERES WAS a ninth-grade dropout who got his start in the working world cleaning toilets at age fourteen. He's in his midtwenties now, with a calm confidence that makes him seem much older. The confidence is only to be expected. Despite, or perhaps because of, his rough start, Zach has become the director of the Michael Polanyi College at the Universidad Francisco Marroquín in Guatemala City.

Zachary Caceres, director of a thriving college of the Universidad Francisco Marroquín in Guatemala City, managed to steer his way through a number of educational obstacles by seizing control of his learning at an early age.

I'm in Antigua, the onetime capital of Guatemala, sitting with Zach in the restaurant 7 Caldos. Zach learned Spanish only when he arrived in Guatemala several years ago, but he casually orders a beer, chatting in Spanish with the waiter as I stare with perplexity at the menu. *Kak ik*, Zach explains, is a strongly flavored turkey soup. *Pepián* is a meaty, spicy stew.

Next door is my hotel, the Casa Santo Domingo, a grand lodging built on the grounds of what was once one of the largest convents in the Americas. The massive stone walls of the convent crumbled and fell in the 1773 Santa Marta earthquake, so walking among the crumbled ruins on the hotel grounds feels like being in Pompeii.

I'm in Guatemala for a conference, but my real mission is to learn about Zach. This isn't as easy as it might seem, even when I'm sitting directly in front of him. It turns out that getting Zach to talk about economics, philosophy, history, or virtually any other subject is easy. Getting Zach to talk about *Zach*, however, is a tougher task.

Zach's father lost his job during the year Zach was born, going from a well-respected, successful engineer-turned-business executive to managing a trailer park in rural Maryland. Zach's school system, like many others, failed to meet the needs of the diverse group of learners within it. The wide economic divide in the county, worsened by a nearby resort town, only made things harder for students and teachers alike. There's

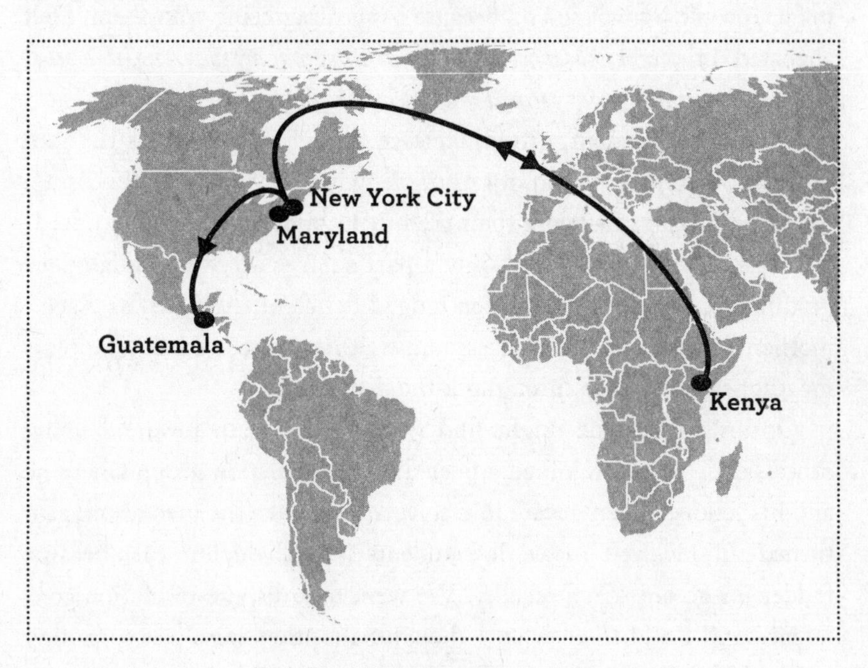

Zach Caceres' learning journey—to date.

finger-pointing aplenty about the cause, but the truth is that some U.S. public school systems are downright awful. To put matters mildly, Zach's school district did not meet his needs.[1]

The negative effects were real and personal. Teachers often showed up late and the students spent a good deal of time unsupervised in trailer "classrooms" with too few desks. Thirteen-year-old Zach and some of his classmates were often miserable, and they inflicted their misery in turn on one another. It was a bit like *Lord of the Flies*—to keep themselves entertained, the kids staged fights, like in *Fight Club*.

Zach was bullied starting from a young age. He was smaller than the other boys, bookish and nerdy. The main issue, for Zach at least, was the school's fear-driven culture, where manipulative and abusive teachers, with a few rare exceptions, made life hell for anyone who thought differently. And Zach definitely thought differently.

"I've always been a contrarian," Zach says, taking a swig of his beer as marimbas play in the background. "Everywhere I went while I was growing up, people were pissed off because I was disagreeing with them. I felt alienated. I thought, *I keep seeing all these things that are so wrong, that other people seem to think are so great. I must be dumb. I must be a bad person.*"

Zach's independent, gifted, creative way of thinking was the root of the problem. As a child, for example, with his swift way of writing, he would produce far more than the other children. But the standardized tests were scanned and only a particular space was available for grading—meaning he was often judged deficient. And yet, as Zach's mother notes, many of his classroom writings were used as exemplars for teacher workshops throughout the district.[2]

Optimistic that he might find *some* positive path toward helping others, early on Zach joined a peer-dispute mediation group where he and his fellow students were to resolve arguments. The mediation, as it turned out, involved uncovering students' issues, which in turn became fodder for gossip. Zach recalls: "We went to a dispute-resolution conference and I told them what I thought: that they were just spreading gossip. It didn't go over well. That evening, a boy ended up throwing ice at me in the hotel room we were all staying at. Next thing I knew, we were in a fight."

A fistfight at a dispute-resolution conference.

Zach joined the Boy Scouts with enthusiasm. For his community service project, he created an after-school music program for elementary school kids, complete with an instrument donation system to receive donations from people in the local community. But as he gradually discovered, his project ran along very different lines from the usual Boy Scout projects, like building and upgrading playgrounds. The projects were evaluated by a panel of parents, one of whom seemed to see Zach as a threat to his own son's success. After all his hard work in getting the musical program ready for launch, Zach was told that teaching and managing a music operation did not show sufficient "leadership capacity" and was thus ineligible as his Eagle Scout project. Disillusioned, Zach left the troop.

Zach's interest in music was to cause him similar problems at church. He was selected to travel to Utah with his youth group to compete at a youth talent convention with a jazz composition that incorporated the song "Amazing Grace." He was told (in the mid-2000s!) that "Jazz has no place in the House of God," and disqualified. Unconventional music wasn't Zach's only problem—his youth pastor eventually took him aside and said, "You have to stop pointing out all the inconsistencies in the things we're saying all the time."

Zach's intelligence, independence, and creativity—all of which would have been admired in another environment—caused him nothing but trouble. As he reached adolescence, his actions took a turn for the worse. With other boys, he would break in to construction sites and smash windows, throw paint on the walls, and steal materials. Together with his "bad boy crew," he threw rocks at cars, egged police cars, and once tried to burn down an abandoned house. He would charm and make out with girls on trips to Christian retreat centers. In short, he was becoming a little asshole—angry at life.

"It makes me sad to think about this time," says Zach. "So many of the people I knew from those days later died from accidents or drug overdoses. They exist now only on my Facebook friend list."

But all of these actions were essentially practice steps at asserting himself, even if his choices just deepened his frustration. Zach's relationship with his parents and family, the true lodestones of his life, frayed. He didn't understand himself or his relationships well enough to realize what was going on.

One difficult day in ninth grade, everything changed.

The woods by the bus stop had become Zach's refuge. He would hide there each day, letting the bus go by. Then he'd head home or simply wander around the neighborhood like the lackadaisical truant that he was. Day by day he grew more miserable—and bolder. One day, his mom was delayed in setting out to work. Zach, who couldn't even muster up the effort to hide, was caught.

The story tumbled out. Zach wasn't just ditching school that day; Zach was skipping school almost every day. And when he did go to class, he was miserable.

That evening when the family sat down for dinner, a television-show-like intervention ensued. Everyone was sitting around with grave expressions talking about Zach's "schooling problem," as if the problem centered only on Zach. One thing became clear: Zach wanted out. The bullying and poor teaching were putting him on a dark path.

The solution he proposed was gutsy—and scary.

➜ Key Mindshift

The Sometimes Lonely Path of the Creative

Sometimes creativity can leave you feeling out of step with those around you. Millions around the world have experienced this "marching to the beat of a different drum" feeling, so if you have periods in your life when these feelings are particularly pronounced, it's nice to know you are not alone.

Finding Self-Sufficiency

Zach would have been fascinated, even as an adolescent, to learn of research that might have shed light on his situation. As it happened, premier sociology researcher Joan McCord had long been involved in a study of at-risk youths. This was the Cambridge-Somerville Youth Study, which was originally conducted in the late 1930s and early 1940s. The study examined how boys go awry in their lives—and how they might be directed onto a better path.

McCord was a vivacious, gifted academic who had herself gone through difficult circumstances. While in grad school, her relationship with her abusive, alcoholic husband had finally imploded in divorce, which had left her as the struggling single parent of two boisterous boys. In the early 1960s, when women were expected to be homemakers rather than breadwinners, McCord's life became a treadmill of grading and teaching to support her children. But she still kept on with her studies, earning her doctorate in sociology from Stanford in 1968. She found herself drawn to research in criminology. The question that kept arising for her was: *How is it that people veer off track in their lives?*[3] McCord's work with the Cambridge-Somerville Youth Study would provide unexpected answers to her question.

Influential American criminologist Joan McCord wasn't afraid to question prevailing wisdom about intervention strategies meant to help high-risk youths.

The study, which was one of the most ambitious programs ever developed to prevent delinquency, was initiated in the 1930s by a researcher named Richard Clark Cabot. It had been carefully designed to quantify what sorts of assistance to youngsters, in the form of counseling, tutoring, and other support, worked best to improve children's lives in the long term.

The study included more than five hundred boys in the Boston area. Both "difficult" (that is, juvenile delinquent) and "average" boys were included. The youngsters were first put into matched pairs controlling, to the extent possible, for family size and structure, neighborhood, income, personality, intelligence, physical strength, and many other characteristics. Then one person from each matched pair was randomly selected for the treatment, with the other going into the control group. The treatment group received a wealth of resources, while the control group received no attention or support at all.

Counselors were assigned to the treatment group. The counselors went with their young charges to athletic events, taught them how to drive, helped them get jobs, and even assisted with family counseling and the care of younger children.[4] Many of the treated boys also received tutoring in academic subjects, were given medical or psychiatric attention, and were sent to summer camps and other community programs. The control group, on the other hand, simply conducted their lives as usual.

In 1949, roughly five years after the experiment ended, researchers traced the subjects of the study. To the researchers' surprise, they found no measurable beneficial effects for the treated group of boys.[5] The obvious conclusion to the researchers? It was too soon to evaluate the program's effects. The investigators felt that the program's benefits would become apparent once the boys were evaluated yet again, after another decade.

In 1957, while still a graduate student, McCord first came into the picture—she was offered a small sum of money to retrace the effects of the experiment on the boys. McCord's work was tedious, and yet, because the Cambridge-Somerville Youth Study had been carried out with exquisite attention to detail, also rewarding. The case records included twice-a-month reports made for more than five years, providing several hundred pages of detail for each boy. After months of painstaking scrutiny, McCord found the same results as previous researchers: none of the expected benefits of the study showed up. There was no difference, for example, in arrest rates, the number of serious crimes committed, or the ages at which crimes were committed. Clearly it was still too soon for researchers to tell whether the boys would experience long-term benefits from the program.[6]

The study's data sat teasingly. Something about the Cambridge-Somerville study kept drawing McCord back, but she couldn't figure out what it was. Had the follow-up studies somehow missed some vital clues? There were small hints that the treatment might indeed have been beneficial, despite the previous "no effect" findings. For one thing, some of the boys themselves, now adults, believed the assistance had been worthwhile.

The National Institutes of Health became intrigued as well. They agreed to provide financial support so that McCord could hire a small team to retrace the study's participants.

Since there had been more than five hundred boys in the then thirty-year-old study, McCord and her team faced the prodigious task of finding the study's original participants and comparing how their lives had unfolded. The team was forced to become amateur investigators, collecting evidence from all angles—city directories, motor-vehicle registrations, marriage and death records, the courts, mental health facilities, and alcoholism treatment centers. Though they were looking for research subjects from a study that had been completed some thirty years before, they located an unbelievable 98 percent of them. Even more amazingly, some 75 percent of the men, then in their late forties and early fifties, responded to the team's questions.

The feedback from the men themselves was straightforward. Two-thirds of the men thought the program had been helpful—they felt the program had "kept them off the streets and out of trouble." They reported that they'd learned how to get along better with others, to have faith and trust in other people, and to overcome prejudices. Some men felt that without their counselors, they would have ended up in a life of crime.

The program should have made a big difference in improving the lives of those who had been in the treatment group. *But it was the opposite.* The treated group was strikingly different.[7] Although the differences were obvious, they were easy to overlook, as they had been in prior reviews of the data, because the effects were so unexpected. Those in the treatment program were more likely to commit crimes, to show evidence of alcoholism or serious mental illness, to die younger, to have more stress-related disease, to have occupations with lower prestige, and to report their work as not satisfying. Not only that, but the longer the boys were in the program and the more intense the treatment, the worse the long-term outcomes became. The program was flat-out detrimental—this was true of both the at-risk kids and the average kids. Another crucial aspect of McCord's study was the finding that the subjective reports by the study participants themselves were unreliable.

Why had the treatment—well meaning and carefully designed in every particular—been so harmful for so many?

Zach's Reboot

Over the course of many heartfelt conversations, it was agreed that Zach would complete the ninth grade and then spend the summer researching alternative programs. With his parents, Zach toured private schools, which were either unaffordable or not a viable alternative to the local public schools.

The more they went around, the more Zach could see only one solution: become a ninth-grade dropout. He told his parents, who didn't take him seriously at first. In the end, Zach persuaded both his parents by explaining that he wasn't going to stop his studying; he was just going to go about his education in a way that would allow him to actually *learn* something instead of sitting there in misery all day in school—in between getting kicked around. There were online programs, he pointed out, that offered a way forward.

Zach still remembers the first day that he didn't have to go to school: "I took a long walk in a nearby forest and had an experience I can only describe as spiritual." He realized at last that he had the freedom to become his geeky, independent self without fear or shame.

Both of Zach's parents worked long hours—they had no time to school him themselves. So they instead gave him rules: he needed to talk to them regularly to show he was learning something, and he needed to somehow hold a job—he wouldn't be allowed to just hide in the house. Occasionally Zach's father would leave questions written on a napkin for him to find in the morning at breakfast.

On the day Zach dropped out, his guidance counselor told him that he was "about to sign away any hope of a good future." His extended family also criticized Zach's choice, telling his parents that they were ruining his life by allowing him to leave school. Since he was no longer enrolled in school, Zach couldn't join the band, do any extracurricular activities at the local high school, use the library, or access college scholarships. It was a difficult time.

But it was a time when all of his previous bad behaviors fell away. The new, more positive outlets he developed for himself diffused his frustration and served as a constructive channel for his energy. His relationship with his parents immediately improved, and he stopped lying about where he was and what he was doing. In the end, Zach has come to believe that dropping out saved his education and is perhaps the most important decision he has ever made. Being out of the constrictive cradle of conventional education and away from the sometimes malign influences of his peers allowed him to begin to find his "authentic" self, even though it did make negotiating with institutions harder because of his lack of conventional diplomas and credentials.

Zach's early, real-world education consisted of holding down a job, using a library card and the Internet, and bringing a lot of curiosity to play. He took some courses online, where he excelled, showing the power of his new learning environment. He became a constant reader—a habit that has served him well ever since in his learning. Zach's outside-the-system educational experience also had the inadvertent effect of putting him on an entrepreneurial path—he repaired electronics he found in dumpsters behind retail stores in order to sell them on eBay.

Zach's erratic, unusual education has left him with one regret: He lacks the solid foundation in math and science that would have made him better able to grapple with technology. But he's still managed to do pretty well. One factor that improved his learning skills was his involvement in music, which became easier as his day became more flexible after he dropped out of conventional schooling.

One day, Zach's father invited him to attend a jazz band concert being given that afternoon at the university. As the concert ended, the professor explained the band was open to all. Zach took the professor at his word and gave him a call. When that call wasn't returned, he called again. And later, again. Eventually, his perseverance paid off—a meeting was scheduled. Zach recalls asking, "What can I do for you if you teach me?"

That was Zach's first apprenticeship. He fell in love with music.

Zach was an early consumer of virtual music lessons—years before Skype became widely used. He even pulled together $100 for a single online video lesson from guitar great Jimmy Bruno.

Learning to play the jazz guitar taught Zach how to be a more "structured geek." Previously, his thinking and studies had often been random and haphazard, but the guitar demanded meticulous thinking. Zach gradually became aware of the importance of procedural fluency.[8] That is, he became aware of the value of a daily practice regimen in creating solid neural patterns that he could call to mind smoothly.[9]

Zach also learned the importance of deliberate practice, in which he focused repeatedly on the most difficult aspects, to help stretch his learning beyond the comfort zone.[10] There are cultural elements in the world of jazz that push musicians toward these important aspects of learning. Zach notes, "If you show up at rehearsal and are playing the same licks, people make fun of you." They call it the "woodshed." As in, *Why are you not in the woodshed? Why are you not practicing?*

When Zach was sixteen, he enrolled at a local college to take for-credit courses with his musician mentor, who was a professor there. A year or so later, he was able to go to New York University (NYU). He officially entered as a college transfer student—which meant nobody scrutinized his high school background.

But in his senior year of college, during a final review of records prior to graduation, he was asked to send a copy of his high school diploma. Of course, he didn't have one—he had earned his 3.98 GPA without the benefit of high school. (He had earned his GPA, incidentally, while working full-time *and* commuting two hours a day.) Once again, Zach found himself frustrated by bureaucracy—he was forced to get a high school diploma retroactively using the University of Texas High School online program.

Zach graduated summa cum laude from NYU with a combined degree in politics, philosophy, and economics. He also was inducted into the Founder's Club, an award for achieving the highest academic bracket offered by the university. He became a research assistant to a historian at NYU and also received a grant to travel across Kenya with a federation of street traders to study informal economies. He became fascinated with entrepreneurship in the developing world.

One day, while working on a business start-up called Radical Social Entrepreneurs in New York, he received an e-mail from a man named Giancarlo Ibárgüen—the president of Guatemala's prestigious Universidad Francisco Marroquín.[11] Giancarlo invited Zach to visit and explore the possibility of collaborating on a number of projects. At age twenty-five, Zach took charge of the Michael Polanyi College within the university and created a radical, profitable experimental program in liberal arts studies. In this program, students design their own degrees.

It's all reminiscent of the way Zach successfully went through his early years. Companies are pounding at the door for the new program's creative, can-do graduates—100 percent are employed or have started their own company. It's clear that Zach's work with the Michael Polanyi College is just a launching pad for greater things to come, both for the college and for him.

Zach loves what he's doing in the developing world. He says, "Third world countries have all the disadvantages that you might expect—poverty and a lack of educational infrastructure." But he also finds something tremendously freeing about working in this arena.

The bottom line is that Zach is highly entrepreneurial. It was hard for him to finally accept that his bent toward business—what was always jokingly described by others as his "side hustle" or his "latest crazy idea"—is actually his vocation. He studied economics in college because it allowed him to look at entrepreneurship and its effects from a big-picture perspective. He felt the usual business classes such as accounting and marketing were creating bureaucratic administrators more than they were training people to accomplish challenging new ventures.

"It's harder to do interesting things or have interesting ideas as an entrepreneur if you have exactly the same body of experiences and knowledge as everyone else," Zach says. "Getting an MBA can be a homogenizing process. Plus, skills like accounting or marketing can be learned as you go. You have to be trained emotionally and psychologically, not just rationally, for entrepreneurship. The abstract nature of so many business classes doesn't train the right habits of mind for building something from nothing, which is often just the unglamorous slog of solving boring little problems each day compounded over time."

Many highly successful entrepreneurs, Zach notes, are not intellectual at all. Because they are not intellectual, they receive intense feedback from reality—not from theories. In point of fact, they sometimes don't have the training—or the working memory—for highly abstract and sophisticated intellectual theories.

"Successful entrepreneurs start doing things like optimizing garbage routes, and, after ten years, they own many of the collection routes in a local area. They've solved a seemingly mundane but still very important problem. Their inability to understand conventional, highly sophisticated concepts allowed them to conceive of something that is extraordinarily useful and, in its own way, highly sophisticated. They become like the world expert of local garbage route optimization."

Zach smiles as he observes: "I know it sounds cheesy, but I do believe that there is genius in all people. Too often education snuffs out our differences, instead of giving people the autonomy to do something great."

McCord Digs Deeper

Like Zach, Joan McCord followed her own internal aptitudes—the research avenues she explored were quite different from those of typical academics. She found it difficult at first to publish her findings that a seemingly helpful family support program was harmful. The program had used many of the approaches that are advocated even today, nearly eighty years later. Submission after submission was rejected. But publish she eventually did. McCord's *American Psychologist* paper, "A Thirty-year Follow-up of Treatment Effects," stirred enormous controversy.[12] It also prompted researchers to begin looking more carefully at well-meaning, superficially beneficial treatment programs. Soon, evidence began to mount of other treatment programs that did more harm than good or, at the very least, did no real good despite sizable expenditures.

There were a number of possible explanations for the poor outcomes that McCord observed. It might have been that intervention by the agency led the boys to be unhealthily dependent on other outside influence. Or possibly, after the boys got used to receiving services, they began to think of themselves as needing help.

Joan McCord's son Geoff Sayre-McCord has followed in his mother's footsteps with academic research—he is the Morehead-Cain Alumni Distinguished Professor of Philosophy and director of the Philosophy, Politics, and Economics Program at the University of North Carolina at Chapel Hill. Sayre-McCord told me: "Mom suspected that an important part of the explanation had to do with the kids adopting over time the (upper-middle-class) norms and values of the counselors, which were not well suited to their lives and prospects."[13]

Joan McCord was a pioneer iconoclast who was willing to question whether well-meaning, beneficial-sounding programs truly met the goal of helping their subjects. McCord found that social programs almost never constructed the procedures needed to reliably evaluate their success. In fact, she found that those working in social interventions often took affront to anyone who wanted to evaluate their programs, since they felt good intentions alone should have been a guarantee of their effectiveness.[14] Program designers regularly avoid collecting the data that would provide valid evidence for the effectiveness of their interventions. Sayre-McCord, who has published extensively on moral theory, meta-ethics, and epistemology, conveys his mother's findings when he tells me: "More often, I think, people completely trust their gut instincts and the subjective reports, over the short term, from those in their program. Of course the Cambridge-Somerville Youth Study shows those are completely unreliable, but it is hard to shake people from their confidence. Also, I think, a lot of people (convinced of the value of what they are offering) think setting up control groups is just failing to help people who could be helped. They would much rather use the money to help more people than to set up some scientific study to confirm what they already 'know.'"

McCord would go on to become the first female president of the American Society of Criminology. She bravely disputed the effectiveness of all sorts of revered helping institutions—boys' clubs, summer camps, young offender prison visits, D.A.R.E. (Drug Abuse Resistance Education), and other popular programs. And she began a process, still slow to take hold in social science, of taking more careful stock of whether a social program actually does what it sets out to do.[15]

MacArthur Award winner Angela Duckworth's life's work has involved advancing our understanding of how to promote gritty, persistent, stick-to-it behavior.[16] Duckworth points to the research of psychologist Robert Eisenberger at the University of Houston, who has found that giving children easy tasks with plenty of rewards *reduces* their industrious stick-to-itiveness.[17] Assistance that makes things too easy, in other words, can backfire and stifle internal drive. The best nurturing for gritty people, Duckworth finds, includes both tough and loving relationships.

When we lift the cover off many programs and institutions, it can be surprising how far their results are from their stated objectives.[18] Programs to train good teachers are themselves perhaps as elusive as genuinely beneficial social programs. Professor of education Lynn Fendler has made the remarkable observation that "there appears to be no conclusive scientific research of any sort that substantiates the effect of any courses in the teacher education curriculum on the quality of teaching."[19] We may want students to find success through conventional pathways, but we need to accept that conventional pathways can be very problematic—sometimes for reasons we don't yet understand. All of this can stifle the spirit of society's most visionary and creative individuals.

Key Mindshift
Avoid Fooling Yourself

As Joan McCord's work shows, sometimes we can feel so *certain* that our approach is correct that we don't examine other possibilities. Part of learning well is being able to remain open to others' ideas and to intentionally work to create situations where we can discover whether we are wrong.

Zach's Mentors

Zach's path is especially interesting because early on, at the end of middle school, he intuited that the ultimate social program—conventional schooling—wasn't working for him. In the end, he chose an unusual, self-directed path that may have given him more of a chance of success than conventional schooling, or many of the off-the-shelf mentoring and guidance programs. Zach's "off the school grid" path wasn't perfect—it was hard for someone in his shoes to get the daily practice that can provide special early-formed skills and ability in math, music, or language. But it was highly successful for him.

Zach credits the role not only of music, but of good mentors and apprenticeships in his life. His first mentor was the music professor. "I did all kinds of menial stuff—cleaned his office. It wasn't glamorous, but it was a way of saying thank you for access to his knowledge." When Zach went to NYU, he helped the economic historian by going to the archives and reading thousands of pages of boring government documents about the financial crisis of New York City in the 1970s and photocopying key documents. "It was really about finding positive sum relationships," Zach notes. "Giving back, not just receiving."

Mentors, Zach feels, give most of their time for free. For Zach, then, the key became how he could make himself valuable to the mentor. He notes: "How am I supporting what my mentor is doing? Because their knowledge rubs off on you through proximity, like osmosis."

Mentoring through the social programs of the Boston study group didn't seem to work, whereas the mentoring Zach received worked well. Zach's intuition is that his mentoring worked precisely *because* it wasn't institutionalized. He wasn't in a mentorship program or organization, and he had no counselors trained to be professional mentors. Instead, the mentorships were relationships that arose spontaneously from searching for opportunities in day-to-day life.

Zach says, "These mentorships were not aimless and around some general sense of 'positive influence' on young people. I was in those relationships because I wanted to learn music, or I wanted to learn economics. I believe this is categorically different than the generic 'positive influence' mentorship because both parties are bringing something to the table.

"My mentors didn't drive me places or give me lots of life advice. They told me things like, 'This is how you analyze a classical composition. Go home and analyze this one and come back next week and show me how you did it.' Or, 'This is the subjective theory of value and why it's important. Go home and read essay XYZ and come back next week and we can talk about it.' They were not my friends, really. It was more like how I imagine a medieval blacksmith's apprentice would operate rather than the interactions of a well-meaning twentieth-century social worker."

Joan McCord's research revealed that social programs are not necessarily a panacea. Zach's own life showed the greater "social program" of a conventional education system sometimes just doesn't fit, either because the conventional system is broken, the youngster is too unconventional to fit in, or both. In the end, a person's own efforts to make their way independently in the world can also lead to a worthwhile, fulfilling life.

What Zach found through both mentoring and his own self-studies was a confidence in himself and his own ability to grapple with tough situations. In a word, grit. And no matter how you slice it, the best person to learn grit from is yourself.[20]

★ **Now You Try!**

Positive Paths to Learning

Zach's life story is inspiring because it reminds us that there isn't a one-size-fits-all formula for education and success. Zach used his learning to get himself off a path of delinquency and onto a more positive path. Finding your own positive path to learning can help improve your outlook in myriad ways. Now is a good time to reflect on your learning pathways, and the goals they lead toward. What are your learning goals? How can you best reach them? Under the title "Learning Goals," put some of your thoughts in writing.

Chapter 6

Singapore

A Future-Ready Nation

PATRICK TAY IS one of the sunniest, most optimistic men I've ever met. But there's more to him than just a positive personality.

Patrick holds two important roles. A lawyer, he is an elected member of Singapore's Parliament, representing the country's West Coast. His other formal role carries the portentous title of assistant secretary-general, director of Legal Services and PME Unit of the National Trades Union Congress (NTUC). (PME stands for "Professional, Managerial, and Executive.") Patrick comes from a hardscrabble, lower-middle-class family—he worked for years as a policeman before joining NTUC in 2002.

Despite Singapore's tiny size—its entire population of 5.5 million people is contained on an island averaging eighteen miles across—understanding Singapore is no easy task. The watery boundaries leave the little city-state like a peremptory end-of-sentence punctuation point at the end of the seven-hundred-mile-long Malaysian peninsula. Singapore contains a diverse population of Chinese, Malays, Indians, and other groups, all knit together by a common understanding of themselves as Singaporeans. Although the language of instruction in schools is English, most Singaporeans are bilingual or trilingual, speaking English, Mandarin, one of the many Chinese dialects, Malay, or the Dravidian language of Tamil.

Singapore is also unusual in that it has no natural resources other than a deep ocean harbor. The city-state doesn't even have enough fresh water for its population. Some precious water is imported across the causeway from the occasionally less-than-friendly Malaysia. More water is obtained through clever, Singapore-developed desalinization processes that are now used worldwide.

In 1965, unemployment in Singapore was in the double digits. Workforce literacy was a mere 57 percent.[1] Stuck in what could have been a cultural backwater, Singapore could have been like many of the other struggling colonies spun off from the British Empire after World War II.

But Singapore turned out differently.

A bustling hive of growth, Singapore now has 2.0 percent unemployment—among the lowest in the world.[2] Its per-person gross domestic product is a sizzling 321 percent of the global average.[3] Singapore's children regularly score among the top students in the world in the PISA—an international assessment of fifteen-year-old students' abilities in math, reading, and scientific literacy.[4] Crime rates are so low that parents feel comfortable allowing their teenage children to gallivant around the heart of the city in the middle of the night. When Singaporean women arrive early to a lunchtime gathering at a local restaurant, they will leave their handbags on tables to mark their spot as they head off to the restrooms. Like many people, Singaporeans do complain about their busy lifestyles and the high cost of living, but their lives are free of many of the ills that people in other nations complain about.

An important part of what Singapore is doing right may relate to how it is approaching learning lifestyles and career resilience. To explore this, and to pick up more of Patrick Tay's insights, I met with Patrick in his offices on the twelfth floor of the thirty-two-story NTUC Centre building in the central business district of Singapore. It's a short walk from the legendary colonial architecture of the highfalutin Raffles Hotel, where each room comes with its own dedicated butler. The mirrored skyscraper of the NTUC Centre sits near the edge of the Singapore River and has a sweeping view across the inlet to the iconic, boat-topped Marina Bay Sands hotel.

Patrick Tay, a member of Singapore's Parliament as well as an influential figure in the National Trades Union Congress, has done much to help Singapore grapple in creative new ways with career resilience. Patrick is shown here in the gym—he is also a black belt in Tae Kwon Do.

Patrick has the upright posture and sturdy build of someone who knows that physical fitness leads to mental fitness. His broad, friendly grin puts me immediately at ease—he starts the conversation by announcing that he is married with three kids. Scholarships propelled him through the university, after which he completed a four-year program in law at the National University of Singapore. The government scholarship requirements meant that he had to then serve the country for six years. He chose the police rather than the more usual role as a prosecutor. During this service, Patrick earned his master of laws, specializing in international law and international business.

However, Patrick is a bit of a crusader—always looking to have a positive impact on the lives of others. He'd kept himself involved in community work and legal outreach, often in an effort to fight for justice for the poor.

After he'd finished his service, his intention was to go into practice as a lawyer. But the NTUC recruited him because he was active in voluntary community work—*and* he had what he calls "deep skilling" as a lawyer.

Patrick joined NTUC in 2002. He shakes his head in disbelief at the rapid passage of time. "Here I am, fourteen years later, still working for the NTUC. There's a lot of satisfaction seeing that what you advocate for comes to fruition and benefits others. Not just five or ten others, but sometimes thousands, or hundreds of thousands. That's what keeps me going day after day."

Patrick's assignments at the NTUC have taken him through a broad variety of industries, including shipbuilding, private security, health care, and now, the financial sector.

"When I retrace my steps," observes Patrick, "it always boils down to advancing the interests and welfare of the workers. Employment and employability seem to crop up day in and day out. We need to bring investment into Singapore—to create good jobs and good-paying jobs. But we also need jobs that cater to our workforce, because our workforce's profile is changing rapidly."

Singapore is a career bellwether for what is happening in much of the developed world. Emphasis on education has led to a workforce weighted toward professionals, managers, and executives. As general demographics have skewed older, so, too, has the labor force. Looming continuously is the specter of job obsolescence. Hard-won techniques, technologies, and even relationship skills can gradually lose their value. People must master new software, different equipment, novel management methods, and even different ways of interacting with others. Traditionally, careers have been stepping-stones where you lingered at each step. Modern careers, however, are more like conveyor belts. You have to keep moving and learning no matter what stage you're at.

Patrick's concern for his constituents shines through as he explains, "We need to redesign our jobs, and we need to 'upskill' people to take on these new jobs. Everyone has to play a role in this—the worker, the employer, the government, and, in the greater scheme of things, society itself."

What binds matters together in Singapore is "tripartism"—a meeting of minds among government, unions, and business employers. "This is key and unique to Singapore," Patrick explains. "Tripartism is not new—it's been in existence for a long time under the international labor organization framework. But I think Singapore has its own unique blend of tripartism. In fact just this morning before I came here, I was having breakfast with our tripartite partners, as we do every Wednesday. We were discussing the same issues you and I are talking about right now. We're one of the few countries that does this—that has employers, government, and unions conversing in the same room. We have one major shared goal, which is to grow the economic cake, rather than try to divide it. We all realize we should not be coming at matters with the sense of who gets the bigger slice or the bigger crumb."

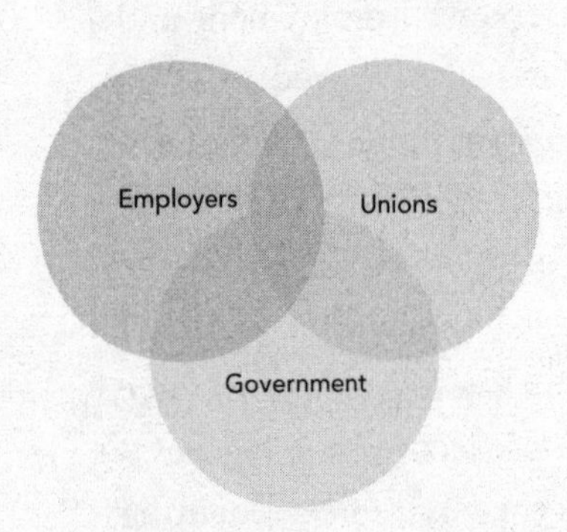

Singapore has an unusual "tripartite" approach where government, unions, and business employers work together to enhance the workforce. Frequent meetings between the disparate groups help them to build personal relationships and find common ground.

Everyone plays a role in this, Patrick notes. By joining together and speaking dispassionately in one room, issues are examined from a variety of perspectives. What does the individual worker need to do? What kind of responsibility does a company have to redesign jobs, automate, innovate, and be more productive? How can the government enable workers to realize their potential? And how can society itself support the shift in terms of social and political models? Singapore well knows that solutions to these questions are vital if it is to stay relevant in the face of an aging, white collar–oriented workforce.

Patrick Tay's secret is that he has simple, yet elegant, solutions.

"T" Versus "π" Approaches to Career Building

Traditionally, career development has been thought of as having a T-shaped trajectory. A person trains to acquire one in-depth area of expertise, be it accounting, mechanical engineering, or twentieth-century British literature. This deep expertise was then balanced by a variety of other, lesser "horizontal" skills—computer abilities, people skills, a hobby in woodworking. Patrick, however, began lobbying several years ago for what he calls a π-shaped approach to career building—two areas of deep knowledge, balanced by a modicum of knowledge and ability in other areas.

> Traditionally, careers have been stepping-stones where you lingered at each step. Modern careers, however, are more like conveyor belts. You have to keep moving and learning no matter what stage you're at.

Traditionally, career development in Singapore, as elsewhere, has been thought of as a "T" shaped trajectory, with one "deep" area of expertise, and many lesser areas of knowledge and interest.

In the new economy, it was clear to Patrick that you should not just have one area of expertise. Even if you went through the trouble to acquire *two* tiny fields of expertise among the millions of areas of expertise available to humans, that's still twice as big as what it had been. Two areas would give far more options and flexibility.

π

Patrick Tay has championed a "π" shaped approach to career building—two areas of deep knowledge, balanced by a modicum of knowledge and ability in other areas. Also known as "second-skilling," this approach to careers builds in resiliency and flexibility in the face of society's rapid growth and change.

Patrick realized that, in a modern economy, "second-skilling" is necessary for career resiliency—it gives you options and flexibility. Naturally, if you already have an intense, difficult-to-acquire deep skill such as being a medical doctor, you can't easily just pop over and pick up another equally difficult-to-acquire second skill—say, becoming a lawyer. But no matter what your first skill, you protect yourself by having some second skill—deeper than just a dabbling in another area. That second skill can either complement the first or give an alternative path if your personal situation changes. Implicit in Patrick's approach is that we can all learn more than we might think.

People often make the mistake of thinking that First World economies such as Singapore's enable the luxury of career change. But this is a misleading perception. Singapore's economy, much like many First World economies, has gone through many peaks and troughs, even in Patrick's lifetime. There was an economic crisis in 1998, then another in 2003 due to the severe acute respiratory syndrome (SARS) epidemic, which cut travel in Asia to skeletal levels. Another hit with the 2008 subprime mortgage crisis.

"With job obsolescence, one deep skill might not be relevant in two or three years' time—things are changing so rapidly," Patrick notes. "There have been retrenchments, downsizings, restructuring, and offshoring. In this new, modern economy, you cannot have just one deep skill. It's good to future-ready yourself with two deep skills.

"For example, people can work in a bank and have an in-depth understanding of a particular niche type of work, or type of software and how to use it. But if that particular financial product or kind of work becomes obsolete or goes offshore, then you'll be out."

I ask whether two skills are something that every worker can have. Could, say, a bank worker have a second skill?

A bank worker *needs* two skills, Patrick explains. In the volatile banking industry, for example, a bank executive can be the first one to get the ax if she doesn't meet her sales targets. A backup skill can be vital. But developing that second skill can be surprisingly straightforward—sometimes there are incipient skills just waiting to be developed.

For example, there is a particular niche that Patrick calls "relationship banking." This type of banker doesn't just have banking skills, they have relationship skills. And skills at relationships are valuable in other areas:

counseling and social work. People in these fields are in high demand in Singapore due to its aging population and other social challenges. If a relationship banker can second-skill him- or herself in, for example, counseling—he or she can hop into the high-demand social service sector. If there's a financial crisis, in other words, there's a fallback.

Singapore funds programs to support second-skilling in both young and old. In fact, those who are forty and above can get enhanced funding when they undergo certification in counseling programs—even if these counseling-related certifications are not relevant to their work. That is, unlike employer-supported programs that offer funding only to train employees in skills relevant to their jobs, the government also funds individual-initiated programs that may not be directly related to their current occupation. The entire country is moving toward individual choice and selection in adult continuing education.

Thanks in part to Patrick's lobbying, Singapore is practical in how it approaches funding. Through the SkillsFuture program, every Singaporean who's above twenty-five years old receives five hundred Singapore dollars in a virtual credit account. This money is then used to offset

I remember being in the midst of hiring a bunch of people at a company I worked for and being presented with an article that said there often wasn't that much difference between someone who'd been working the same job six months versus six years. Second-skilling needn't be as difficult as many people think. Skill development curves are typically logarithmic, not linear. This means that while developing deep expertise may take a long time, you can often rapidly accelerate to the point of diminishing returns in a fairly short period of time. And this is often good enough to get a toehold in a new area. Personally I find I enjoy acquiring many skills because of the thrill of the initial rush of progress.

—Brian Brookshire,
online marketing specialist at Brookshire Enterprises

training expenses in anything they might want, not just what their company wishes. "You might think five hundred dollars is not a large amount," says Patrick. "However, many programs are already funded 80 to 90 percent. So the five hundred dollars can be used to pay for the unfunded portions, which, previously, we had to fork out from our own pockets."

Why support second-skilling from an individual's, rather than an employer's, interests? This encourages the employee to build on their skills: to up-skill, re-skill, multi-skill, and second-skill—and to give employers the funding to incentivize this process.

Key Mindshift
Second-Skilling

Second-skilling is a good idea in today's swiftly changing career environment. A second skill can allow you to be more nimble if the unexpected arises in your day job.

Practicality, Passion—or the Allure of Money?

Patrick explains that second-skilling has two dimensions. The first is the work dimension. Here, your second skills can allow you to move *into* or *across* or *up*, either for career advancement or because you've lost your job. Your second skill may also arise because of your passion or interest.

An IT friend of Patrick's, for example, has a great love of visual design and graphics. Although he does backroom technical IT support in his regular job, he went for training in 3-D design and graphic design. So now, although he's still in IT, he does lucrative media and design freelance work on the side.

"So you have the work angle, and the passion angle," Patrick explains. Of course, if you can marry the two, that's ideal.

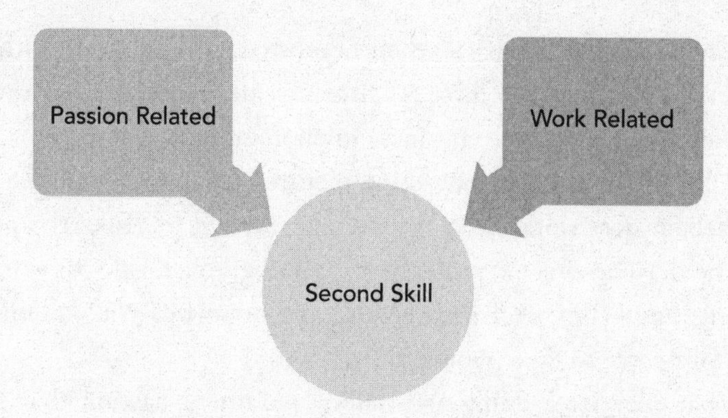

A second skill can grow from work-related needs, or from passion. An optimal second skill can arise from both.

For the work angle, it can be best to look at job trends and forecasts to see where hiring will be occurring. In Singapore, growth areas for the next five to ten years include advanced manufacturing, health care, and aerospace. (More straightforward manufacturing in Singapore is being offshored to China and other parts of the world where it can be done cheaply.) The aging population is causing an increase in the demand for health-care services, and the country is trying to build an aerospace hub.

What about a typical mechanical engineer? I ask. How would he or she second-skill?

"Because the engineering mind is logical and process-oriented, an engineer might second-skill in any of those high-demand areas," Patrick says. "So if you are an engineer with an expertise in tunneling and underground mining, you might actually be able, with a little second-skill training, to improve supply chain quality in health care."

Of course, the time when second-skilling becomes important is often also the time when you're starting a family. How can people handle second-skilling when time is tight? Patrick gave me two examples of friends of his who'd done this. Both had a passion for photography.

Patrick's police officer friend included his family in his hobby. He took beautiful photographs and videos of his children in action and received encouraging feedback when he posted on Facebook. Although he'd served on the police force for fifteen years, he decided to leave his career and start a freelance business in photography.

An IT friend also started taking pictures for fun and left the computer sector after working in it for some seven years. Like the policeman, he has become a professional photographer, shooting events, weddings, still-lifes, and nature. He now organizes trips around the world for people to take photographs.

"What they were doing as a hobby became a passion that transformed their careers," says Patrick. "It's the same even in my own life." Patrick used to conduct workshops about employment law, labor legislation, and industrial relations in some of the undergraduate programs in local universities purely for fun. He's also a certified swimming instructor and Tae Kwon Do instructor.

I venture to Patrick that he doesn't take a π-shaped approach to his career—it's more comb-shaped. But I get what Patrick is saying—that, especially if time and money are tight, you should try to build your second skill out of what you are already familiar with. You often have more talent and ability within you than you think. Second-skilling isn't necessarily about a job—it's also about respecting your multifaceted ability to be good at different things.

Mushrooms, Stacks, and More

Geometric approaches to thinking about careers can be useful. Another possibility, beyond T and π, is a mushroom approach—with a big fat stem as well as a broad umbrella. U.S. sales entrepreneur Rodney Grim, for example, keeps his focus on his broad profession in selling electronics, but also keeps competency in many related aspects of the business, jumping from one area to another as he sees opportunities. Rodney has been a marine technician, then a land mobile phone technician, then he worked in land mobile sales and then for a manufacturer. Now, however, in running his own company, he has clients in many different industries, and his "secondary" programming skills often come in handy.

There are other shapes beyond the "T" and "π" approaches that can be used to help envision careers. A "mushroom" approach means having broad competencies, all supported by a broad stem of expertise.

This is akin to the "talent stack" approach outlined by humorist Scott Adams of *Dilbert* fame.[5] It isn't that Adams is a virtuoso of many talents. It's that he has a combination of many often mediocre abilities that combine to create a formidable talent stack. As he himself describes it, Adams is a second-rate artist with reasonable writing, business, marketing, and social media skills. Put all these middling skills together, however, and it becomes more clear as to why Adams might be successful as a cartoonist.

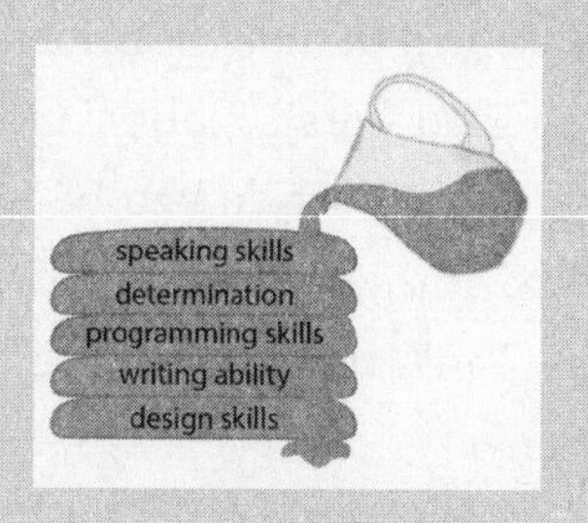

Scott Adams of Dilbert *fame describes the "talent stack" approach to understanding career success. It's easy to forget that there is much more to a successful career than expertise in one specific area.*

Many individuals focus on acquiring a specific skill—say, a certain programming language—but forget that other skills, such as being able to speak humorously and effectively, can add formidable value to their talent stack.

Though Patrick understands and appreciates the value of earning a nice living, one of his pet peeves is people who sniff out their career choice based simply on the highest possible pay. "People often see the banking and finance industry as a sexy profession because they see these people driving big cars—Ferraris, Lamborghinis—and living jet-setting lifestyles." He notes that the banking industry is air-conditioned—a big deal in sultry Singapore—and you get to golf and wine and dine.

However, it turns out that this sector is not as rosy as many think. "I would say maybe one in one thousand make it to drive a Lamborghini and a Ferrari," Patrick says. People are expected to meet weekly, monthly, and quarterly key performance indicators, and "the exit is always there. It's a harsh environment."

"Half of the people who graduate from the National University of Singapore in engineering don't eventually become engineers," Patrick observes. "Because their training makes big data management and managing data analytics easier, engineers can get a cushy job in the banking and finance industry. They have a swanky office with a high starting salary, much higher than if they were to become an engineer. A lifestyle of the rich and famous, so to speak. But it's never as cushy as they think."

The problem here—as so often is the case in a number of careers—is that rosy expectations often run head-on into harsh reality. On the other hand, by simple definition, not everyone in a field can be in the top 2 percent. And not everyone needs to be at the top to find that field fulfilling.

Preventing Early Career Misfires

When you're twenty-two, can you really decide the entire course of the rest of your life? Careers sometimes go off track simply because people have to make their career choices at an early age. It's easy to think delaying the choice of career until a person is older will solve the problems. But instead, such delays can result in a host of other problems, particularly when a career requires lengthy training.

For this reason, Singapore is addressing the issue of career choice and is bringing in career counseling early on—as early as possible. This way, students can be much more informed about the realities and requirements of the areas they are dreaming about. Singapore's approach includes "learning journeys," internships and vocational attachments that begin at a very young age. Youth group programs such as nEbO introduce younger people in upper secondary school and higher levels to specific companies and sectors. Patrick explains that this helps people "avoid a rude awakening in a challenging sector," minimizing the expectation gap between school and career.

However, no matter how career-related learning is approached, there are always trade-offs. For example, sixteen-year-olds are mighty young to be locked into one career pathway. Still, without knowledge of what a given career is really like, they are even more likely to be dissatisfied when they emerge from college.

Understanding this, I asked myself: Is it better to keep students open to broad career paths as long as possible in their schooling? Or do you encourage them to lock into certain areas that—early on—seem to be more suitable to them?

To answer these questions, I took a trip down the block to the halls of government in Singapore.

A Big-Picture Perspective

Dr. Soon Joo Gog is the chief research officer and a group director at SkillsFuture Singapore, a statutory board under the Ministry of Education. She is a slender, focused dynamo of a woman who has thought deeply about how learning can and should take place—not only from an individual perspective, but through governmental policies that support prosperity. Gog (pronounced *Gokh*, rhyming with the Scottish word *Loch*) has worked hard to foster a desire for lifelong learning in Singapore's varied peoples.

The three-million-strong Singaporean workforce is too big for Gog and her team to reach in its entirety. Instead, the team works in conjunction with employers, trade associations and chambers, trade unions, and education and training providers such as universities and vocational institutions to build capacity for change through learning.

Dr. Gog's petite stature and youthful looks belie her prodigious intellect. As chief research officer and a group director for SkillsFuture Singapore Agency, she spearheads Singapore's continuing efforts to build a learning ecosystem where people are able to empower themselves. Gaining an appreciation of lifelong learning is part of this process. Behind Dr. Gog is the new Singapore LifeLong Learning Institute, with different floors providing training for areas as dissimilar as retail, early childhood, and information, communication, and technology. Singapore pours enormous resources into facilitating adult learning and career resilience.

"This capacity for change is critical," Gog notes, "because change is the only constant we will be seeing in the future—from technology to economy to social and political structures. Change is accelerating, and we need to build the capacity for change to continue to be relevant."

Careers define our identity in life. But just following your passion isn't enough. Aspirations must be matched with opportunity.

—Soon Joo Gog

Matching Aspirations with Opportunity

"Careers define our identity in life," observes Dr. Gog. At the same time, she understands that the traditional approach of just following your passion about what career you want to pursue isn't enough. "Aspirations," she notes, "must be matched with opportunity."

Some of Gog's work with her education and training partners is about providing career signposts—information that allows people to connect with employers and make transitions from where they are to where they want to go. To this end, the SkillsFuture Agency and Workforce Singapore Agency have been instrumental in building a guidance system for individuals and employers to discover labor information and access a jobs bank, skills profiles, and course directories. The system will cater to people at all stages of their careers to support them in finding opportunities for learning and for new potential jobs. Using government-initiated institutions like e2i (the Employment and Employability Institute) and programs like CaliberLink, people can connect with full-time employability coaches. These coaches do career counseling when people lose jobs or are looking to move up to or across to new jobs. The Jobs Bank allows employers to search for the most suitable candidates.

A pot of tea cools untouched before us as Dr. Gog and I reflect on our own careers, both of which emerged from a sort of accidental opportunism—who we happened to meet and what we happened to read from the limited books and magazines available to us back when we were making career decisions. The Internet has changed all that.

Gog marveled at how much more information there is for current job seekers. If you love music, you can much more easily see what it's like to be a composer, a performer, or a high-quality sound technician. Not so much is left to chance—we can learn from the experiences of others with the click of a mouse.

In Dr. Gog's estimation, some 80 percent of people can self-navigate their way through the educational system and into their careers. But people who reach career inflection points where they're fired or laid off are sometimes devastated—they feel all doors are closed. Part of the problem, Gog observes, is their own mind-set: "People often think they can only do whatever they did in the past. But if people are given insight into the many opportunities they can explore, they aren't as likely to feel so helpless and angry."

Singapore's approach to career capital and resilience might be likened, as Dr. Gog's colleague May May Ng later points out, not to a safety net, but to a springboard. "Safety nets may sometimes be helpful, but they can also serve as traps. The approach we take is more like a trampoline. People may have to bounce down as they gather and prepare themselves, but ultimately, using their own strength, they can bounce high."

Because Singapore is a small country, the government attempts to keep the focus on high-growth industries. Pharmaceutical research and development is important, as are such fields as logistics, freight, information, communication, and technology. Other areas include network security and software coding, tourism, health care, social services, and education.

Statistically, changing jobs in Singapore seems easy—after all, the unemployment rate hovers around 2.0 percent. But this can be misleading. In low-wage positions such as those in retail, busing tables, and so forth, where the skill requirement is minimal, changing jobs is straightforward. But the more expertise a position requires, the trickier job change gets. In certain sectors, such as engineering, employers are looking for relevant job experience. This limits who can apply.

> Safety nets may sometimes be helpful, but they can also serve as traps. The approach we take is more like a trampoline.
>
> —May May Ng, manager at the SkillsFuture Singapore Agency

The Singaporean approach is not just about having a vibrant economy—it's about having and creating good jobs. Gog observes: "A good job is not just about pay. It's about autonomy to make decisions to improve work and accessibility to upgrading skills. It is about professional identity."

Gog has a sense of excitement about what's ahead for Singaporeans. She reflects on the economist Joseph Schumpeter and how the workforce development system can encourage people to benefit from the "creative destruction" of the economy. "Our job at the agency is to help the whole system to evolve. It's not just about vocational educational systems or about a university. It's about creating and nourishing a skills ecosystem where people can empower themselves."

"Career capital" plays a role in all this. Cal Newport, a Georgetown

University computer science professor and the author of *So Good They Can't Ignore You: Why Skills Trump Passion in the Quest for Work You Love*, has observed that "Career capital are the skills you have that are both rare and valuable and that can be used as leverage in defining your career."[6]

But Gog takes this further, explaining, "Sometimes you can't say you're learning for either work or leisure, because you never know when it will come in handy. Like Steve Jobs with his training in calligraphy and typography—he never knew this would be part of the unique feature of Apple, where the fonts are always so beautiful."

Key Mindshift

Significant Change Is Possible

It's easy to fall into a rut of thinking that you can only do what you've done in the past. But enormous change and growth is possible if you open your mind to the potential.

Self-Empowerment in a Cradle of Equal Opportunity

Critical to Singapore's approach is the empowerment of individuals to ensure that everyone has an equal opportunity to be successful. This sounds idealistic, but as a small country, Singapore is able to efficiently coordinate key stakeholders, from schools to parents to communities to employers to industries.

Dr. Gog smiles as she reflects on her son's recent project for school, and then points out: "People sometimes make the mistake of thinking that Singaporean children's high performance on the PISA examination is reflective of simple rote learning. But the PISA is actually a test of problem-solving skills—it's no rote drill. In Singapore, children don't just learn facts and subject matter. They are exposed to what might be called 'deep thinking skills' at virtually every step of their education.

Literature, for example, is about analytical skills—analyzing context and situation. Kids are asked to figure out what a story is trying to explain to us at a deeper level. At a foundational level, math is about problem solving. The kids are asked how logical thinking is used, how we inquire. Literature, math, science—it's never just about the subject. It's also about what the subject teaches about interacting with life at a deep level."

Gog explains that a learning system must be looked at more broadly than as just a school system. It needs to include family and community. Learning takes place in the context of an entire country and culture, and every country has its own social economic pact—how it defines parental involvement.

Skills are of strategic importance for Singapore. This means institutions do not decide alone which skillsets, curricula, and teaching methods are suitable to drive and support business growth. The Singapore education system isn't treated as static, but is rather continually refreshed. Colleges and universities often partner with businesses to identify emerging skillsets so that students and graduates always obtain the latest and most useful and important skills. This is not to say that the arts and humanities are neglected in Singapore. Indeed, the multilingual heritage woven into the fabric of the country guarantees an appreciation for multiple perspectives—an appreciation that more ethnically homogeneous countries can often only dream of for their students.

In Singapore, education at lower levels is free, and at the university level, it's 75 percent subsidized, with lots of scholarships. But financial support is only part of the picture. Parental support for education is generally strong—and Gog feels this is key. There is also a built-in societal expectation for people to work hard.

Gog reflects on other systems: "In the U.S., it's quite different, because there, states and cities have a lot of say in the education. There are no standard practices. Some cities are extremely successful—some aren't. It's hard to transform a failing school because you need to transform the entire community."

Singapore systematizes ways to support the small percentage of students who encounter major challenges in school, such as a serious illness or the death of their parents or a learning disability. NorthLight School, for example, takes in the students from various schools around Singapore who have failed the "primary school leaving exam" more than twice. NorthLight teachers embrace creative learning methods that build the confidence and passion of the students. In one such classroom example, teachers ask their students to flip the card in front of them from green to red if they don't understand. Positive tones and positive reinforcement also help include parents. NorthLight also has a work-study program with job coaches who help ease some learning-disabled students through vocational studies and into the workplace.

"Education is never just about school. It's about how to build up the ecosystem," notes Dr. Gog. "We're trying to ensure everybody has a pathway up."

Key Mindshift

Fostering a Learning Lifestyle

A learning lifestyle is something that can be nurtured and grown in communities, nations, and cultures.

Learning at Large

This idea of inclusiveness as a key aspect of learning isn't just lip service. SkillsFuture works intensively with partners and the community to show how lifelong learning can be a part of their lives. The weekend after my visit with Dr. Gog, there was a lifelong learning festival, created to nurture a mind-set that people can learn for leisure and for work—anytime, anywhere.

As in many countries, Singapore's education system is not perfect. Anecdotally, some Singaporeans point toward their educational system as being part of the reason they are often not as creative as Westerners. Other Singaporeans quietly observe that they are good at taking exams and memorizing and solving problems but not so good at "thinking outside the box" and finding new solutions.[7]

It is possible that Singapore's educational systems, like other test-intensive systems around the world, may stifle creativity, perhaps because they don't give their most creative students the extra skills they need to overcome certain learning handicaps that sometimes accompany creative thinking.

What is unquestionable is that Singapore is taking active steps to address critical issues, such as an overemphasis on high-stakes testing to determine a person's life and career. Recently, Singapore announced the new Committee on Future Economy, helmed by Singapore's best and brightest. It is geared toward leading Singapore through a future that is "VUCA"—volatile, uncertain, complex, and ambiguous.[8]

"Singapore is always a work in progress," Dr. Gog explains. "We never think of ourselves as having arrived. Once we know we are always a work in progress, we will try to do the next best thing, again and again.

"Singapore," Dr. Gog concludes, "is a learning nation."

 **Now You Try!**

Broaden Your Learning Toolkit

Singapore has a unique approach to encourage continuous learning and second-skilling. How might you apply some of these insights to maintain an attitude of continuous learning? Do you already have a second skill? If not, in what area might you choose to develop a second skill? What might you do to broaden your learning toolkit? Under the title "Skill Broadening," put your thoughts into writing.

Chapter 7

Leveling the Educational Playing Field

AT AGE NINE Adam Khoo was expelled from primary school for fighting. In secondary school, Adam's grades were stuck in the bottom tier. He couldn't pay attention very easily in class.

People who did poorly in school sometimes say they are actually very smart—it was just that the material was somehow boring and beneath them. Adam doesn't say that. Learning was just plain hard for him. Reading books was even worse than listening to a teacher—books just made him sleepy.

It didn't help that Adam's parents divorced when he was in his teens, and he suddenly found himself with a stepsister in the gifted student program. She attended the top school in Singapore, while Adam was in the bottom one. Over and over he was told, "Why can't you be like Vanessa? Get straight A's like Vanessa?"

A lot has changed since then. I met with the forty-one-year-old Adam Khoo at his offices in downtown Singapore. Adam is now the multimillionaire founder and executive chairman of one of the largest educational training firms in Southeast Asia: the Adam Khoo Learning Technologies Group. Though Adam has a reputation as a titan of industry, he's also a nice guy who hopes to inspire others by sharing insights about his early struggles and what had shifted his path.

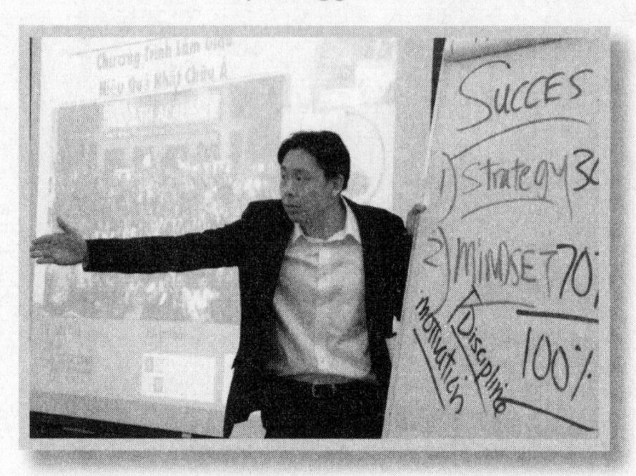

At age thirteen, Adam Khoo swiveled from incorrigible failure to astonishing success by reframing his mind-set and acquiring powerful new techniques for learning. He now leads a company devoted to helping people make similar successful changes in their lives.

I enjoy traveling to Singapore, because its common use of English makes it such an easy place for Westerners to navigate. Adam's native language, for example, is English—he's had his share of struggles trying to pick up Mandarin.

However, one aspect of Singapore that can be difficult for Westerners to wrap their minds around is the level of competition. High population densities in places like Singapore and elsewhere in the East mean that, whatever you are doing, you can find yourself competing with hundreds of thousands or even millions of other people—often with goals identical to yours.

Perceived success in Asia, as in many other parts of the world, is often tied to material success—a job with good social standing, a high salary, and education. Parents place a lot of pressure on their children to become doctors or lawyers, despite the fact that not everyone is geared for those disciplines, and the world doesn't need only doctors and lawyers.

Math and science are considered particularly important for long-term success. Parents urge their children to excel in math so that they can go to a top university and have their pick of lucrative careers. Singapore has been working hard to change attitudes—to show that artists and athletes, for example, can be just as valuable as engineers. But old mind-sets are slow to change, particularly when people look at economic realities. The bottom line is that it's often harder to make a living at more "fun" jobs like being a writer or musician.

School competition is especially intense. Standardized tests are the equivalent of lining up several hundred thousand students at a starting line, firing a pistol into the air, and giving medals to the few who struggle to a nanosecond lead across the finish line. Because of this, school preparation has turned into an arms race. Students are being drilled at younger and younger ages. It used to be that if a student scored four A's on Cambridge A-level exams—the big exams that dictate which universities you might enroll in—they were among the best. Now, it takes seven A's.

Singapore is working to reduce the pressure such tests put on students and to change the tests themselves to foster more open approaches to thinking and learning, but the system is still punishingly rigorous. However, in other Asian countries less attuned to the need for change than Singapore, the testing systems can be even more brutal. Millions of students compete for just a few slots in top universities. Those with inferior scores are relegated to lower-prestige regional universities, colleges, and vocational schools. Many simply aren't able to progress academically after secondary school.

"Good" students in Singapore, as elsewhere in Asia, are often channeled in early grade levels into faster tracks. This makes sense—it keeps students moving at a pace that's more suited for them. But in Asia's face-oriented culture, it *matters* how your child is classified. This puts a double burden on students. If you do poorly on a test, it isn't just that you yourself have failed. You've also failed your family—shaming them in front of others. And there's other baggage: If you're put in the slowest group, you're also among the rowdier and less academically driven kids. This makes it harder to concentrate and reinforces feelings of inadequacy. You also lose access to the best teachers. Your thoughts become home to the refrain: *I'll never be as good as an upper-tier kid.*

Once you're on a downward scholastic path, it can seem impossible to change. Everything is geared to drive you further down.

But it *is* possible to reverse course—even if you're one of the "slow" kids.

 **Now You Try!**

Mind Tricks for Success

Have you ever felt outclassed by intense competition? As this chapter will reveal, there are mind tricks that can help you get back in the running. Can you anticipate what some of those tricks might be? Under the title "Mind Tricks," put your ideas on paper.

Adam Khoo: A Reboot on Life

Adam Khoo was a latchkey kid—one of tens of thousands in Singapore who shared similar circumstances. When Adam came home from school, there was no one to even check on him. This was perfect as far as he was concerned—all he cared about was playing computer games. He had no interest in school, which was frustrating for his parents, who spent a lot of money on tuition and tutors.

Adam drove the tutors away, running around and ignoring what they were trying to teach him. Failing grades in school? He didn't care. Beyond computer games—along with comic books and television—he was only interested in hanging out with friends and getting into fights. Adam was an attention seeker—always wanting to stand out. If he couldn't stand out in a good way, he was happy to stand out in a bad one, picking arguments and joining disruptive groups of other wayward youths.

Adam's businessman father was loving, but he was at a loss for how to encourage and motivate his son. Adam's mother was one of the top journalists in Singapore—a high-flying career woman. She was also loving, but unhelpful in a different way. When Adam found himself unable to do math, she would shake her head and say, "I think you got my genes."

"I had this label placed on me; I was lazy and stupid," says Adam. The labels weren't that far off base—Adam is the first to admit that he's a slow learner, and being slow made him want to avoid his studies. What Adam hadn't realized is that there are mental tricks that could have allowed him to overcome the drawbacks that arose from his "nonstandard" intellect. As his success has since shown, once he overcame those stumbling blocks, he could leverage his brain's advantages.

It's trendy to think that motivational youth camp programs don't make any real difference in a student's life, but for Adam, at age thirteen back in 1987, one of these programs did. It was Super-Teen Camp—the first such program in Asia.

"I was exposed to ideas from the human potential movement," says Adam. "*We are all born gifted. There is no failure, there are only learning experiences.* I was inspired by examples of people who had turned their lives around."

Adam reflexively taps the desk in front of him as he bounces back and forth in his chair. He has a flickering kind of attention—his smarts don't necessarily build in solid, step-by-step blocks of conventional rationality. It's no wonder conformist schooling nearly called it quits on him.

"In the Super-Teen program, I learned tricks with memory. How to visualize and associate. I would say to my friends, 'Give me fifty words. I'll memorize them in five minutes. Let's bet two dollars.' I got a new identity—as a genius. But it was really all in the techniques I'd learned."

Adam was a creative individual who loved daydreaming, drawing, cartooning, and music. Staring at the pages of a book was boring for him. But a technique called "mind mapping" he'd learned in camp allowed him to pick what was in a textbook and reorganize it in a highly visual way, using cartoons that triggered his memory.

But the program didn't just teach learning tools. It also taught him to dream big. The instructor challenged him, "Do you want to be mediocre or exceptional?"

"Exceptional."

"If you want to be exceptional," the instructor said, "you have to set goals that you cannot achieve with your current level of capability. You need goals that will make you stretch."

This made sense to Adam.

He aimed high—very, very high. He set a goal that year, when he was thirteen, to go to Victoria Junior College, one of the top junior colleges in Singapore. The secondary school Adam was attending wasn't the best—in fact, no one had ever gotten into Victoria from his school. His teachers told him the goal was unrealistic.

"That was one of the hardest things for me—the people I was around," Adam says. When people heard his goals, they made snide remarks: *You're crazy. Can't be done.* To his dream of entering the National University of Singapore, his father just said, "There's no way." But Adam is by nature a contrarian. The more people would tell him he couldn't do it, the more he'd set his mind to it.

But on top of all this, Adam had another dream: to build his own business. He built this dream by visualizing and imagining. In secondary school, he would draw his own name cards. *Adam Khoo, Chairman.* He developed a dream that he could *be* someone—that dream kept him motivated. He pasted his goals all over his room. *Junior college, here I come! National University of Singapore, here I come!* A poster that says, *I'm a winner.*

Adam started topping his class. His geography teacher couldn't understand what was happening to her former "bad boy" student, but she knew enough to take advantage of a good thing. She asked Adam to use a class period to teach his friends what he was doing.

Adam started to write articles that he printed and distributed to friends: how to set goals, manage time, stay motivated. The other students began to look up to him, to respect him. He gained a new identity, and he enjoyed it. That was when he discovered his passion: *He was here to inspire.*

To deal with the fact that he was a slow learner who was always behind in class, Adam started asking his teachers what they were going to be teaching the next day. He'd preread the chapter and do a mind map.

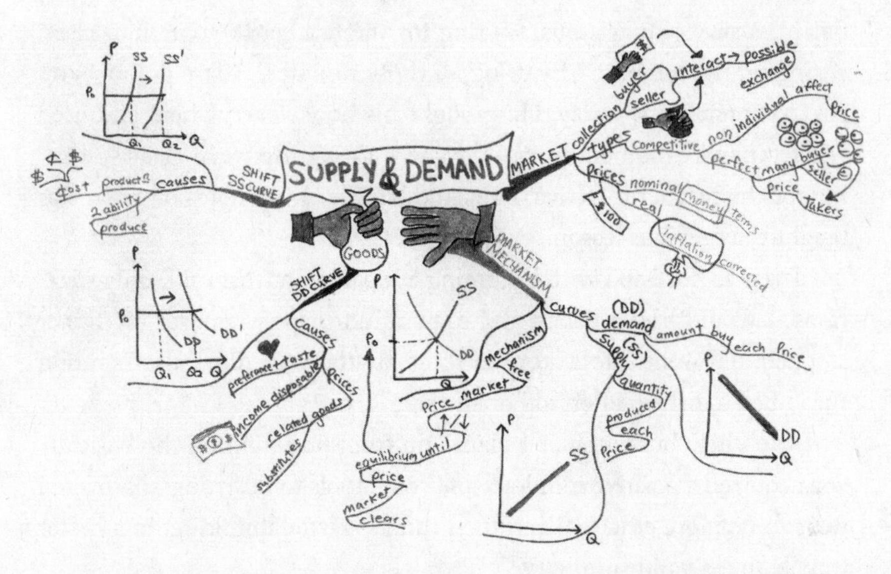

One of Adam's mind maps from college. Drawing out the key concepts of what he was learning about helped him remember and understand what he was learning. Notice the drawings embedded throughout Adam's mind map—sketches like these have been shown by research to enhance your ability to learn and remember the relevant concepts.[1]

The next day, listening to his teacher, he would be hearing the information for the second time, which meant he understood it better. He would ask a lot of questions in class, which allowed him to add to his mind maps. He drew funny cartoons that helped him remember.

He worked hard, especially in math, which was not a natural subject for him. In junior college, he majored in math, despite, or perhaps because of, its difficulty. Mandarin was also difficult for him. At that time, it was virtually impossible to advance without it. Adam spent half his time studying the sounds and characters of a language that he just couldn't seem to grasp. He failed important tests time and again over many months. Finally, he got a D. "Suddenly, there was a glimmer of hope!" Adam grins. "And then I kept studying, took the test again, and got a C. I got to go to junior college!"

In the meantime, he DJed and performed magic, becoming a very efficient time manager. There are many pockets of time people ordinarily waste—riding a bus, waiting for the teacher to come into class, sitting on the toilet. . . . By using all those minutes, Adam got an extra two to three hours a day. He brought his books everywhere and used whatever spare moment he could to be learning. Between classes, waiting for the teacher to arrive, he would be tidying up notes on what was taught the previous lesson.

"I was so obsessed with becoming a top student that on family vacations, I would find a bench and draw mind maps whenever my father stopped and went into a store to shop. By the way, did I also mention that I had a girlfriend on top of all that?"

I can't help but laugh as I glance up from note-taking. "So basically you acquired a positive mind-set and some tools for learning, discovered ways to be more efficient, and then things started unfolding in a better way. Is that a valid summary?"

"Yes. But it was lot of hard work. When I started secondary school, for example, my math teacher called my mom up and said, 'Your son should not even have passed primary school and come to secondary school, because his level of mathematics is so low.'" But because Adam was so motivated to get A's in secondary school, he went back to his primary school books and started practicing every question to really learn the basic concepts. It was hard work—not like some kind of easy miracle. Adam's stepsister could read a new chapter of math and get it pretty much instantly, whereas Adam struggled, reading and rereading until whatever he was trying to grasp finally made sense.

Adam knew how easily he could think he knew a question, but then forget or make a careless mistake when it came time for the exam. "So even though I knew how to do a question," he says, "I'd cover the answer and do it again. Cover the answer and do it again." Adam would keep at it until the ability to do the problem somehow became so fluid that it was subconscious.

Five Key Mindshifts That Enhance Learning

1. Create vivid mind-map sketches that bring the material to life.
2. Memorize by using visual association.
3. Use pockets of time that are often neglected, such as sitting on a bus.
4. Practice over and over again until you can work a tough problem with ease.
5. Visualize your successful future with whatever you are learning.

What Lies Behind Good Learning?

Adam's "do it till it's subconscious" approach is backed by sound neuroscience. The article "Reading and Doing Arithmetic Nonconsciously" by psychologist Asael Sklar and colleagues, published in the *Proceedings of the National Academy of Sciences of the USA*, opened plenty of eyes about the fact that multistep, effortful arithmetic equations can be solved below people's level of awareness.[2]

A technique known as Flash Anzan teaches children to rapidly add numbers on a mental abacus. And by "rapidly," we're talking *spooky*-fast—numbers in the hundreds or even the thousands are flashed momentarily on a screen. The number 3,492, for example, is quickly replaced by 9,647, then 1,785, and so on, as youngsters watch and do their mental summation.[3] Kids enjoy this technique, which they start learning by drumming on the table with their fingers as if an abacus were there in front of them.[4] Gradually, students learn to let their hands lie still as their minds race ahead.

It's difficult for an unpracticed individual to glimpse fifteen three-digit numbers flashed on the screen in a total of less than two seconds—yes, two seconds—and imagine that anyone could add them that fast. But with practice, it's absolutely possible. And doing this has great benefits. The mental abacus method teaches students to use the visual and motor parts of their brain in adding—a thought process very different from those that use simple pencil-and-paper techniques.[5] Children who use the mental abacus can gain such fluency with mental mathematical processes that they can even do Flash Anzan while simultaneously playing the language game *shiritori*, which involves saying a word that begins with the final kana character of the previous word. It seems the verbally oriented *shiritori* and the math of Flash Anzan use different parts of the brain.[6]

Procedural fluency is a sort of automaticity of thought that arises because you've done something many times before. Examples include being able to casually back up a car (not so easy the first time you might try it!), perform a pirouette in dance, flawlessly repeat a tongue twister, or play a piano concerto. In math, it might encompass the ability to easily multiply two numbers together or, on more advanced levels, to take a derivative in calculus.

Practice is what builds the well-connected neural networks that underlie procedural fluency. You can draw these previously constructed, well-connected neural networks as "chunks" easily to mind when you need to do something that's difficult.[7] These well-integrated mental chunks underlie the partly or sometimes entirely nonconscious ways of thinking that allow people to draw mental patterns easily into working memory. A neural chunk is a little like a computer subroutine—you call it to mind when you need it, but you don't really have to think about what it's doing.

Researcher Anders Ericsson has for many decades studied the development of expertise.[8] He has found that "deliberate practice"—practice that focuses intensely on the most-difficult-to-master aspects of the material—is what moves people ahead most quickly when they are trying to learn anything new or trying to get better at a task they already know well.

Let's take the simple task of tying your shoelaces. When you are first learning to do this, you must concentrate intently, using your working memory. Later, the task becomes so easy and natural that you can tie your shoelaces while, for example, simultaneously telling a complicated joke. All you have to do is think "tie my shoes" and off the subconscious parts of the brain go to do the shoe tying while your working memory focuses on telling the joke. These sorts of practiced chunks can make our lives much easier. If you've ever watched a master knitter or crocheter deftly creating an intricate pattern on a sweater while holding a casual conversation, you've seen the benefits of chunked expertise.

Cognitive load theory, first developed in the late 1980s and increasingly supported by recent neuroimaging research, posits that if you overload working memory, the brain simply can't process the information.[9] As a person gradually develops expertise in any topic or area, neural imaging shows that the areas of the brain associated with working memory seem to calm down, showing reduced activity.[10] Basically, it seems that chunking—those solid, well-connected neural patterns developed through practice and procedural fluency—really does seem to offload thinking processes from the working memory area (centered in the prefrontal cortex) to other parts of the brain. This leaves working memory with a lighter load, giving it space to handle new thoughts and concepts.

Neural chunks developed through procedural fluency may be particularly useful for people who have less capacious working memory. The more you're able to offload tasks onto subconscious, automatic processing through chunking, the more working memory you leave available for problem solving—or telling jokes.

➔ Key Mindshift

Develop Neural Chunks Through Deliberate Practice

Whenever you are trying to learn a difficult new topic or skill, focus on deliberate practice with the toughest parts of the material. Break whatever you are learning into tiny chunks—a little part of a song on the piano, a word or verb conjugation in Spanish, a side kick in Tae Kwon Do, or a homework solution in trigonometry. Practice that chunk of material until you have created a solid "neural chunk"—a pattern that you can easily call to mind and accomplish. Once that chunk is mastered, however, don't fall into the trap of repracticing it just because it's easy and feels good—instead, keep the bulk of your focus and practice on what you find most difficult.

Creating Luck

At age eighteen, the end of Adam's "prepping" time at Victoria Junior College, he had to take his A-level examinations, which determined whether he could get into a university. His dream, of course, was the National University of Singapore. Mandarin came back to haunt him—he again failed it on his A-level exams. But, mirabile dictu, Adam did so well in everything else that he received a special probationary entry. His family was astonished. He was elated.

And then the hard work began. In the end, using the techniques he'd learned and taught about how to grasp and retain difficult concepts, he was to graduate with honors with a degree in business administration. It was a far cry from the failing student he'd been a little more than a decade before.

I wondered how much luck played into Adam's success.

Adam told me that he believes there are two kinds of luck.[11] There's dumb luck—"what we call 'ass luck' here in Singapore"—and then there's created luck. Adam's thoughts are perhaps rooted in astrological beliefs, which can reign strong in Asia. Some people are believed to have "astrological luck" while others are less fortunate.

Adam recalls, "I once had this employee who had that astrological kind of luck. He won two cars in two lotteries in Singapore. *Cars in two lotteries!* He won the weekly lottery so many times it was off the charts, statistically. You just can't explain it. Anyway, when I was young I went to one of these fortune-tellers for fun. He read my astrological path and said, 'You just don't have the luck.' I don't know if this is a self-fulfilling prophecy, but for some reason all my life, whenever I play a game of chance like poker, blackjack, I never win." Adam shakes his head, nonplussed. "I always lose. I don't know why. Maybe there is a bit to this ass luck thing, but I don't have it and I don't need it."

Adam instead quotes the Roman philosopher Seneca: "Luck is what happens when preparation meets opportunity." In order to be lucky, Adam explains, you must have three things.

First, you must have the opportunity. Adam believes that *opportunities never arise looking like opportunities.* They always come disguised as problems. It takes a certain kind of mind-set to turn problems, which we all encounter every day, into opportunities. "Lucky" people are those who see opportunities when others see problems.

Adam laughs. "I have so many opportunities because I keep seeing so many problems!"

The second thing is preparation. "Even if you have an opportunity, if you're not prepared with the right skills and knowledge, you won't be able to take advantage. Like the Boy Scouts say: 'Be prepared.' Constantly make sure that you learn and you upgrade your skills, so that when an opportunity comes along you can take advantage of it."

The third thing is action. "You suffer from paralysis of analysis if you only think without doing. If you don't jump in, you never get lucky."

Adam "jumped in" right after his mandatory army service, teaming up with Patrick Cheo, a whip-smart friend he'd made at the National University of Singapore, to continue his mobile DJ business. Patrick was the operations manager, while Adam served as the DJ and magician. But Adam wanted to find a way to give back to society as well as run a business. One day, he went back to his old alma mater, Victoria Junior College, and told the principal the story of how he'd gone from a failure to success. He asked if he might train Victoria Junior College students in some of his techniques.

The principal agreed, and Adam started doing that. "At first, I didn't charge any money. I was doing it for fun because I loved it. And after a while I found that not only did I love it, but hey, I could make a career out of this! I began coming up with one-, two-, three-day training programs."

It was then that he wrote the book *I Am Gifted, So Are You!* "That book started everything," Adam says. He explains that it could be seen as luck that he had that book published, but what really led to it was having a problem. "The problem was, I was a lousy student who had difficulties learning."

That problem became an opportunity. He realized that he could show that "if a loser like me can do it, so can you." Adam was prepared to write a self-help book because not only had he had an opportunity because of the failure he'd overcome—he'd also read many self-help books. He knew the genre. And he was able to go to the third thing: He took action.

At the time, people questioned what credentials he had to write a book. But he just went ahead and wrote it—four hundred pages. He says, "I went to over a dozen publishers with the manuscript. Simon & Schuster. Prentice Hall. Addison-Wesley. They all rejected me." He just kept resubmitting.

One day, he got a call from the Singapore office of Oxford University Press. They had him come in for a meeting. The editor found his book interesting—said it had potential—but called his English "commercially unviable."

Adam laughs: "Essentially, that meant my writing sucked." The editor told him they'd consider the book if he'd rewrite it—which he did. Next, she edited the content. Adam shaved it to about two hundred pages and kept working at it. His mom helped.

The book was published Adam's second year at the National University of Singapore. "I was so excited!" But he couldn't find it in stores. Adam discovered there was no marketing budget—the publisher couldn't afford to put many resources into a book from an unknown first-time author.

"I thought, *Okay, I'll handle it.*" He went around to the schools and bookshops in Singapore and did free talks. Doing these free talks forced him to learn public speaking. Six months down the road, his book became one of Singapore's bestsellers. It was on the bestsellers' list consecutively for years.

Adam Khoo's Suggestions for Career Resiliency

Adam suggests reading books and taking courses and seminars to keep yourself prepared, no matter what twists or turns your career may take. "The only way to ensure your skills don't become obsolete is to always keep learning," he notes.

Learn more about your area of expertise, but also learn about things outside your area of expertise. Be open to learning even nonacademic topics, just as Adam learned magic and DJing. These two skills, which are completely unrelated to Adam's college degree in mathematics, have contributed a lot to his career by teaching him how to effectively engage with audiences.

Bad Traits as Best Traits

One of Adam Khoo's most admirable characteristics is that he doesn't try to paper himself over publicly as some misunderstood prodigy who deigns to climb off his mountaintop to share his genius. If Adam *is* special—and I believe he is—a part of that is his willingness to share the challenges that his sometimes not-ready-for-prime-time brain has forced him to overcome.

Over the next week while I'm in Southeast Asia, Adam and I are scheduled to speak at some of the same events. As we prepare in the greenroom for an audience of two thousand in Jakarta, I nervously ask if he ever gets stage fright. He gently replies that he used to, but he finds focusing on the audience and their needs gets him out of his own head and past feelings of stage fright. Because Adam is open about his shortcomings, I can't resist asking him about his worst traits.

He comes right out and tells me that he's "not very smart"—and has no problem telling people that. They typically think he's saying that for effect, but he means what he says.

"I have to make things simple so that I can understand them," he explains. But this becomes a positive—it turns out that people like Adam's books because he can make things simple.

"What are some other bad traits?"

"I'm stubborn. A regular donkey. Contrarian. And I'm very naive. My wife and Patrick [his good friend and the company CEO] always tell me that I get conned all the time. So when it comes to negotiation they don't let me be there, because I will give away the farm.

"Patrick's the numbers guy. I'm the creative guy. So we are a great fit. He's the polar opposite of me—very details oriented. I'm the big-picture guy—the dreamer. His catchphrase might as well be 'Adam, stop dreaming.'"

"Other bad traits?" I ask.

"I'm obsessive—a compulsive worrier." Adam doesn't like worrying all the time, but he senses how useful it is for him—he's afraid that if he reduces his worrying, he may not be as sharp. He tends to think of the worst possible thing that can happen, then he prepares compulsively until he feels a confidence arising that he's prepared enough.

"The bad thing about being obsessive," I venture, "is that when something goes wrong, it's hard to stop obsessing about it. That leads to a negative mind-set."

Adam nods. "That used to happen to me." Again, he learned mind tricks—how to reframe and put the issue behind him. He's learned tricks to shift—to know when to stay obsessed and then when to disassociate.

Adam has a wealth of other creative mental tricks to keep him, and his students, on task. For example, he tells his students that motivation is like bathing—it doesn't last.

"You can't wash once and be clean for the rest of your life," he says. "Because no matter how much you bathe, you get dirty and smelly and you've got to bathe again. Likewise, no matter how motivated you are, the world can be a negative place. Things don't go your way. You get criticized. You get 'dirty' again. So you have to learn how to motivate yourself every day, just like washing."

In true tough-love fashion, when Adam runs programs for the kids, he tells them, "I'm not here to bathe you; I'm here to give you the soap, the brushes. You learn how to bathe yourself."

One of Adam's most powerful tricks is reframing. He makes a practice of seeing problems as opportunities—seeing how to transform a liability into an asset. He loved Steve Jobs' reframing of his firing from Apple. As Jobs put it: "Getting fired from Apple was the best thing that could have ever happened to me. The heaviness of being successful was replaced by the lightness of being a beginner again."[12]

Adam also chooses to believe that everything happens for a reason. No matter how bad things might seem, there is something he can learn from it. Remembering this keeps him motivated through setbacks.

For example, when publishers rejected him over and over again, he would tell himself, "This means I must rewrite the book in the way that makes it even more powerful and compelling. When it finally becomes a bestseller, I will have a great story to tell." He did something similar when he was posted to a lower-ranked school, telling himself it was a blessing because he could more easily get to the top there.

Adam says: "I keep playing these crazy movies in my head like I did in the past. I'd see myself on stage performing magic or inspiring people. Constantly playing that movie creates the compelling drive to want to go do that."

Though Adam makes his way through a good bit of his personal and business life using his imagination, he is a constructive fantasist. He's learned to keep playing his goals in his head to make them as vividly real as he can. He feels it's important to review not just what you want to do, but *why you want to do it* every day. And he does that—reminding himself that he wants to be successful so he can help people. Of course along the way, 70 percent doesn't go according to plan, which can be frustrating. To keep himself motivated, he watches inspiring videos on YouTube of people who have overcome great challenges, and he loves reading biographies. "I often see my problems are trivial by comparison."

Nothing works for everyone. But Adam's continuing search for new ways to not only learn what he's actively studying—hedge funds, for example—but to reframe adversity in a positive way forms part of his intellectual strength.

The Power of Mental Tricks

Again, the intuitions that underlie Adam Khoo's thinking and approach are backed by solid neuroscience. A recent neuroimaging meta-analysis analyzed what is termed "cognitive reappraisal of emotion"—that is, reframing.[13] The study revealed that finding positive ways to think about a negative occurrence extinguishes negative emotions arising from the fight-or-flight center of the amygdala. For example, an alarming picture of someone bleeding can be cognitively reframed as "that's just a movie, and they're using ketchup." Or negative feelings about an illness can be reframed into something more positive by focusing on how that person will get better. Reframing is such a powerful approach that it lies at the heart of the cognitive behavioral therapies used to treat depression, anxiety, and other psychological challenges. We'll go deeper into reframing—that is, understanding the context with which we view the world and everything in it—in the next chapter. (Caution, snakes!)

One might think that, in some cases, such reframing would simply be unworkable mental trickery. After all, what if we're talking about real life, and a person you are close to has a terminal illness and they're

Three years ago I had difficulties finding freelance IT assignments. I realized this was actually a positive. It meant it was time to upgrade my skills—to both broaden and deepen what I know. Today, I'm over fifty and have no issues finding work—contrary to most people my age.

—Ronny De Winter,
freelance software engineer, Belgium

not going to get better? In that case, you may want a different way of reframing, perhaps by focusing on the *quality* of life rather than the absolute *quantity*. (Hospice workers are masters of this type of reframing.) It seems consciously finding a way to *change the meaning* of what is experienced reduces the flood of stress-related neurotransmitters released by the hypervigilant amygdala. This provides a pathway for the mind to uncover deeper truths, even if it's through what appears to be mental sleight of hand.

 **Now You Try!**

Making Your Luck

"Lucky" people are those who learn to see opportunities when others see problems. Adam Khoo has achieved a great deal by cultivating an "everything happens so I can learn something from it" type of attitude to reframe adversity.

Bring to mind a major challenge in your life that you can reframe as an opportunity. Under the title "Becoming Lucky," jot down what concrete steps you could take that might allow you to take advantage, either now or in the future, of the opportunity and others like it. On a paper, or better yet, on the back pages of a daily journal that you keep to record your progress and feelings, start making a list of mental tricks you can use to reframe various challenges.

Creativity

In Chapter 3, marketing expert Ali Naqvi alluded to the brain's two fundamentally different operating modes: "focused" and "diffuse." Research has shown that the "focused" mode pops into gear as soon as you turn your attention onto something. The "diffuse" mode, on the other hand, pops up when you're not thinking about anything in particular—

like when you're standing in the shower, looking out the window of a bus, or going for a run. We can't generally be in both modes at the same time—our brain puts its energy into either one mode or the other.[14]

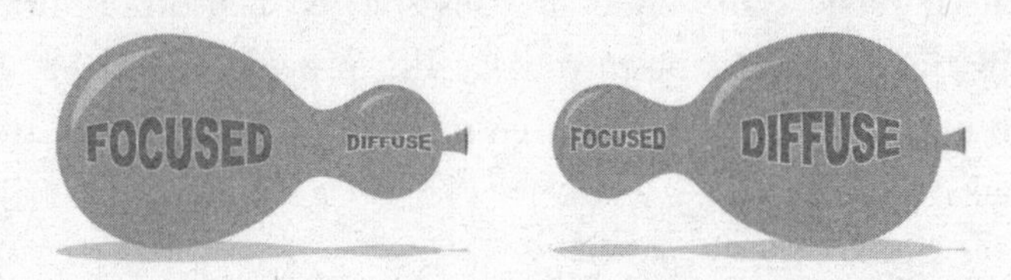

The balloon on the left represents your brain in focused mode—most of the energy is poured into your intense focus. The balloon on the right shows your brain in diffuse mode—most of your energy is poured into the other, more relaxed, diffuse networks.

The diffuse mode is actually a set of neural "resting states," that is, more wide-ranging patterns that our mind falls into when we're not focusing tightly on a task.[15] Creative new ideas seem to emerge from these broader diffuse states.[16] We can slip into the diffuse mode for minutes or even hours at a time when we're daydreaming. But the diffuse mode can also pop up momentarily—even blinking appears to activate it.[17] (Clever martial arts experts look for when their opponent blinks—that micromoment's change of awareness is a good time to launch an unexpected move.)[18]

Researchers are beginning to understand that learning seems to involve two steps. First, you focus your attention, activating your "task positive" networks—you're conscious of this part of the learning process. Then, you take your focus *off* what you're trying to learn by going into diffuse mode. You're not really conscious of this second learning step—in fact, it might seem like you're doing nothing. But this second step is what allows the brain to creatively consolidate what you're learning.[19] Metaphorically speaking, it's as if your brain first focuses to pick up the material in front of you, then as soon as you relax and let your mind wander *off* the material, the brain is freed up to tuck the material away. This is why that little break after your mental workout with a Pomodoro session is so important—it starts giving your brain a chance to consolidate the material you've learned.

It's easy to imagine that educational systems that constantly rely on focused attention could inadvertently inhibit development of diffuse networks.[20] Your brain needs breaks.[21] The effects of too much focusing time might be even further magnified when society promotes relaxation mechanisms, such as certain forms of meditation, that also encourage focus.

In fact, meditation can have surprisingly different effects depending on the type. The vast majority of meditation techniques center on the development of *focused attention*.[22] In contrast, *open monitoring* types of meditation, such as Vipassana and mindfulness, appear to improve diffuse, imaginative thinking. There is a lot of variety in how open monitoring meditation techniques are taught, however, and focused concentration may play a role, or at least a prelude, to mastery of the form.

In the end, practices that encourage focusing can be a boon for learning. But having some daily time when your mind is encouraged to relax and wander freely is also important, particularly if you want to encourage creativity. From a practical standpoint, then, if you are a meditator, you might try to avoid feeling you should *always* be steering your thoughts back into focus if you catch your mind wandering outside meditation sessions.

This might be why people also find the Pomodoro technique so useful in promoting creative productivity. It trains your ability to focus, but the reward when you're done is that you get to let your mind wander onto whatever it feels like. With this technique, it's as if you first complete a focused workout in your mental gym, after which you head to the mental spa—overall an enjoyable experience.

Back now to Adam. It's interesting to note Adam's care in reframing. His occasional obsessive worry involves mind wandering about bad things that could happen—as neurolinguist Julie Sedivy has noted, mind wandering has "a possible connection to neuroticism."[23] Still,

Adam doesn't try to completely eliminate mind wandering—at least, not until after his mind wandering has done its duty in preparing him. He simply waits for the right moment to reframe his thinking.

Working Memory and Mind Wandering

Working memory basically amounts to how much information you can hold temporarily in mind—like the five names of people in a group you've been introduced to. (Wait, was the first one Jack?) Working memory, as it turns out, has a counterintuitive relationship with both intelligence and creativity.

Intelligence is often equated to the strength of your working memory.[24] People with steel-trap minds—strong working memories—often have the enviable ability to hold many aspects of a problem in their head at once. This leads to easier problem solving. A person with a limited working memory, on the other hand, must find a way to simplify complex topics to work with them. The process of finding ways to simplify can be tedious and time-consuming. But surprisingly, research has shown there's a hidden benefit—people with less capable working memories are *more likely* to see shortcuts and have conceptual breakthroughs. It seems the "smarter" person with large and retentive working memory sometimes has less incentive to see the material in newer, simpler ways.[25]

There's another drawback to having a steel-trap mind. If you can understand something by easily holding ten steps in mind at once, your tendency is to explain it to others in ten-step fashion—even if people have become lost after step three. Steel-trap brilliance, in other words, can make it harder to teach others, especially if your steel-trap mind is coupled with a "doesn't suffer fools gladly" mentality. In Adam's case, he's got a teaching advantage. As he notes, once he comes to the point where *he* understands something, he can generally explain it so *anybody* can understand it. There are other advantages to having a limited working memory. The ideas you want so badly to hold in mind may float away despite your best efforts—to be replaced randomly by other ideas, thoughts, and sensations. This may sound less than ideal, but this is also what underpins creativity.[26] Poor working memory, incidentally, is often correlated with attention deficit disorder, so if this condition is making school tougher for you, it's important to realize it also gives you advantages.[27]

You may argue that a strong working memory not only helps with problem solving—it helps as well with getting good grades. But research has shown that there is a countercorrelation between school grades and creativity.[28] The better your grades, in other words, sometimes means the worse your creativity. There is also a correlation between disagreeableness and creativity.[29] It may simply be that disagreeable people are more willing to be brats—to throw aside the compliant, deferential demeanor of their more agreeable peers. Looking back at Adam's wild streak as a youngster, it's possible that it was just a manifestation of his creativity.

Incidentally, enhancing your working memory can be tough. Exercises to build working memory may strengthen your ability to do that particular task, but they often don't seem to build the overall capacity of working memory itself.[30] Only one set of programs, those by BrainHQ, seems to reliably increase working memory.[31] This program won't turn you into a genius, but it does appear to mildly improve memory, processing speed, and general cognition, in some sense stopping or reversing the mental clock as you age. We'll discuss BrainHQ further in Chapter 8.

Whether the effect is small or large, there appears to be a surprising side benefit from such exercises: They seem to improve mood—decreasing feelings of anger, depression, and fatigue.[32] Rather than dampening the sometimes crabby amygdala, these exercises reduce activity in the anger-modulating apparatus of the insula. This is a part of the brain that allows us to experience not only pain, but a number of the basic emotions, including anger, fear, disgust, and happiness. Working memory-related practice may give us stronger mental "muscles" for cognitively managing emotional stimuli. Since learning frequently entails exercises like those used to enhance working memory, this may help explain why adopting a learning lifestyle can just plain make us feel better.

Key Mindshift

The Hidden Good Side of a Poor Working Memory

When you might be struggling to hold something difficult in mind as you are trying to understand it, remind yourself that your struggles may well arise in tandem with your creativity. You wouldn't want to trade your creative streak, even if it does mean that sometimes you need to work a little harder!

Adam Khoo's Helpful "Bad" Characteristics

Before Adam's mindshift at age thirteen, he had plenty of time to play at whatever he wanted because he didn't care about his grades. Later, however, after attending the educational boot camp, he didn't just devote himself to conventional schooling—he also kept up his DJing, magic, and coaching activities. So Adam was always learning and growing—just not necessarily through a conventional academic lens of focused, intense study.

We tend to emphasize obvious positives such as a good memory and an ability to focus as being the most important elements for learning. But sometimes our negative characteristics can also be surprisingly worthwhile. Here's a recap of Adam's "worst," yet most helpful characteristics in learning:

- **He's not very intelligent.** Like many "less intelligent" individuals, Adam seems to have a poor working memory. But his poor working memory forces Adam to simplify concepts and to focus on the central aspects of any situation. It can take him longer to figure things out, but in the end, he understands clearly, deeply, and simply. His poor working memory forces him to discover simpler ways to grasp concepts, ways that are often missed by the seemingly more intelligent. Adam's comfort and familiarity with workarounds have also opened him up to searching for other mental techniques to allow him to grapple successfully with learning—and life.

- **He's a worrier.** Adam has learned to take advantage of his anxiety, using it as a reminder to prepare carefully. Once he is sufficiently prepared, he lets go by reframing his thoughts to calm his amygdala, which allows his anxiety to subside. This approach is akin to applying the serenity prayer—Adam has learned to change the things he can and accept the things he can't.
- **He's a contrarian.** Adam's stubborn nature means that negative comments strengthen, rather than weaken, his resolve to achieve the goals he sets for himself.
- **He's a naive dreamer.** "Dreaming big" has supported Adam in founding his own successful businesses. Adam partners with and listens to practical people to ensure he grounds his dreams in reality.

★ **Now You Try!**

Revisiting Mind Tricks for Success

At the beginning of this chapter, you anticipated some of the mind tricks for success that we would explore. Revisit your thoughts now and add any mind tricks that you might have also learned about in this chapter. Your final list will form a useful reference tool for you in the future.

What Adam Teaches Us

There may well be a reason I feel so comfortable around Adam and his group. Creative people sometimes say that one of the best ways to be creative yourself is to be around creative people. Adam and Patrick are themselves highly creative and they seek highly creative, can-do people for their team. Time and time again, I discover that one of their team members did not do well on standardized tests when they were in school, but they are, say, a world-class computer gamer, or a novelist, or a magician.

It's possible that the high bar set by the stringent Singaporean system of educational testing—similar to testing systems in many Asian countries—has systematically served to select for and reward students with strong working memories. *But the more creative types often have poorer working memories.* In other words, Singapore's educational system doesn't necessarily instill a lack of creativity. It instead may actively select *against* creativity, regularly penalizing creative people for the seemingly less efficient way their brains work. And it doesn't just leave the creative types behind—it leaves them with feelings of hopelessness and inferiority.

What Adam and his group have done is show that certain mental tricks can allow unconventional minds to compete more effectively. This, in turn, levels the educational playing field and allows different ways of thinking to flourish—ways that not only support learning, but that also help improve creativity.

Educational workshops such as those promoted by Adam Khoo are sometimes thought to simply be heightening the education arms race. But there's another way to see Adam's work. He's busy democratizing education: bringing a set of mental skills to great swaths of students—often more creative types, who are frequently washed out of the educational system and discouraged by well-meaning teachers, friends, and family who don't understand that conventional learning tools don't work well for these students. Adam's old identity as an education system reject who found acceptance with outcasts—coupled with his very real leadership abilities and desire to stand out—could have taken him down a dark path. Instead, he forged a positive way forward that allows others to follow in his wake.

Of course, school success is not the be-all and end-all for every student—nor should it be. However, academic success, not to mention general success in life, shouldn't be a zero-sum game with prescribed winners and losers. All of society benefits when a large percentage of the population is both well educated *and* creative. This is still the case when "well educated" just means graduating from secondary school with a decent ability in reading, writing, and working with figures, and being "creative" simply means being able to flexibly envision new approaches.[33]

Adam Khoo was lucky. He came from a pragmatic but loving family with the resources and will to keep trying until they found an approach that clicked. And Adam himself had just the right combination of stubborn, neurotic, naive, optimistic, and creative characteristics to seize the initiative and sally forth—once he grasped the nuances of a new mind-set.

No teaching method comes with a 100 percent guarantee of success. But there are hundreds of millions of students with inattentive minds around the world who don't come from well-to-do families, and who, for various reasons, are unable to peel off the label of academic loser. These students often wash out of educational systems, with their creative abilities untapped. They can be left feeling both hopeless and worthless.

Perhaps it's time for educational systems worldwide to co-opt Adam Khoo's thinking and methods. Both conventional and unconventional students can learn new techniques to become successful learners and to live happy, fruitful lives.

In the meantime, we can all use Adam's insights to our advantage.

★ Now You Try!

Considering What Underpins Your Mindshift

Early on, if your parents were upward striving, they might have encouraged you to follow time-honored paths toward success, such as becoming a medical doctor. Such parental pushes are understandable. Healing sick people, for example, is a darned handy skill, it pays well, and people respect you—not to mention that your career success reflects nicely back on your parents. Some cultures around the world place special value on traditionally successful careers, which means some children can feel stronger pressure to fulfill parental dreams.

But of course, not everyone can be, or wants to be, a doctor.

Your friends have a different set of needs—they often want to see you smiling *right now*, reality be damned. If you want to be a movie star or champion basketball player, your friends will often be right there for you, pushing you forward no matter how unrealistic your dreams might be. This can be why we sometimes see a shocked singer on talent shows who finds herself ridiculed when she finally faces the public instead of her friends. But it's also important to realize that friends aren't always necessarily supportive. Because they see you as part of their world, they can subtly undercut any efforts you might make that would take you out of their circle. And if you are successful, jealousy can sometimes rear its head.

Teachers and professors can provide useful insight on careers, but like parents and friends, they, too, can have their vested interests. A bioengineering professor, for example, may encourage you to become a student in his department (which keeps his department alive) by praising bioengineering as the fastest growing field in engineering. He may not mention that the reason the field is growing so fast is that it started from a tiny base, and that there aren't a lot of jobs for bioengineers.

If you are married, you will have spousal considerations. If you have or desire children, there are other factors in play.

You might think career testing would give direction in your mindshift, but such testing often provides mechanistic feedback about what your strengths and passions are *now*, with little consideration about how you can change.

And speaking of passions, we're often encouraged to follow them. But a world of passion could be an unhappy place—who would build cars or houses, or stock grocery stores, if everyone simply followed their passion?

It's worth noting that mindshift successes can result from a blend of a person's own "pie in the sky" desires with the constraints of the real world. Scientist Santiago Ramón y Cajal, for example, reluctantly became a doctor instead of an artist because of his father's adamant insistence. But in the end, as a doctor, Ramón y Cajal won the Nobel Prize in part because he brought his artist's insight into his work.

In the face of all these conflicting considerations, what do *you* believe? Under the title "Attitudes and Influences on My Mindshift," take the time to reflect and write on the following questions:

- Do you feel that people have a "true potential" that others should support no matter what?
- Should others' considerations be taken into account when you are planning a significant mindshift? If so, how strongly?
- Should the reality of the working world be a factor in your mindshift? If so, how strongly?
- Do you have a weakness you can change into a strength? How can you accomplish this?

Chapter 8

Avoiding Career Ruts and Dead Ends

TERRENCE SEJNOWSKI HAS a broad forehead, a craggy smile, and a quick way with a quip. His lean, athletic build masks the fact he's in his late sixties.[1] You wouldn't recognize him walking the palm tree–lined streets or jogging on the beach near La Jolla, California. Little known even to his neighbors is the fact that Terry Sejnowski is one of the handful of people who are simultaneously members of all three of the top national scientific, medical, and engineering academies of the United States. In the rarefied world of neuroscience, he's something of a legend.

But back in the psychedelic 1960s, when Terry was in his twenties, he was just another student—an admittedly good student, of course, as well as a smart guy. Unfortunately, he was not smart enough to recognize that sitting in on some biology lectures would lose him his girlfriend.

Terry grew up in Cleveland, Ohio, already a science nerd in elementary school. By high school, he was running the radio club. The adviser, Mike Stimac, inspired his students to think big with projects like Moonbounce, which used a commercial radio transmitter and an array of antennas perched on top of the high school to send signals to the moon and back. Stimac, a pivotal mentor in Terry's life, advised the aviation club as well, where Terry learned to fly.

Looking back, Terry notes, "Being a good student and being smart aren't necessarily what's important to be successful. It was at the radio club where I learned how to build things, to have goals, how to plan long-term projects. As the president of the club, I learned how to manage people and work with them toward a goal. It was not formal academics that formed my future career. It was really how to take the knowledge that you've learned and turn it into a new direction."[2]

After earning his bachelor's degree in physics in 1968 from Case Western Reserve University, Terry received a National Science Foundation fellowship and went to Princeton University to study theoretical physics. John Wheeler, the legendary Manhattan Project physicist who coined the term "black hole" and shepherded studies of general relativity, took Terry under his wing as his master's student.

Terry Sejnowski reading Vladimir Nabokov's Ada *as a physics grad student at Princeton, roughly 1976. Note the then-typically Princetonian tie and slacks—this would eventually get Terry into trouble.*

Wheeler was another exceptional mentor, who also encouraged Terry to think expansively. One day, Terry wondered, "What would happen if we have a black hole that's pea-sized?" Wheeler's response? "Terry, that's a crazy idea. But it's just not quite crazy *enough*." No, wedging an object the size of a solar system into a quarter teaspoon just wasn't weird enough for Wheeler.

Getting Around the Barrier of Intelligence

Terry wasn't just focused on physics—he was a people watcher as well. Studying at Princeton, he saw clearly that there were a *lot* of smart people in the world. Intelligence was necessary for his line of work, but he began to realize that it wasn't enough. "Intelligence can, in fact, be a liability," Terry observes. "With intelligence, you see the options, but you also see the barriers. This means that the smarter you are, the easier it can be to talk yourself out of things." One idea he'd wanted to investigate not long after arriving at Princeton—what would a giant black hole look like at the center of a galaxy?—was shot down by some of the professors around him. The topic was later published by someone else—with fanfare. "Persistence is also key," Terry adds.

Terry's time with Wheeler gave ever-deepening insights into physics' most complex challenges. But Wheeler taught him something else: "Terry, everybody makes mistakes. But when you do make a mistake, don't persist in it. Get off that road and get past it as quickly as you can."

The advice would resonate in big ways.

Terry was already immersed in physics, but the subject would eventually occupy virtually every aspect of his consciousness. At the time,

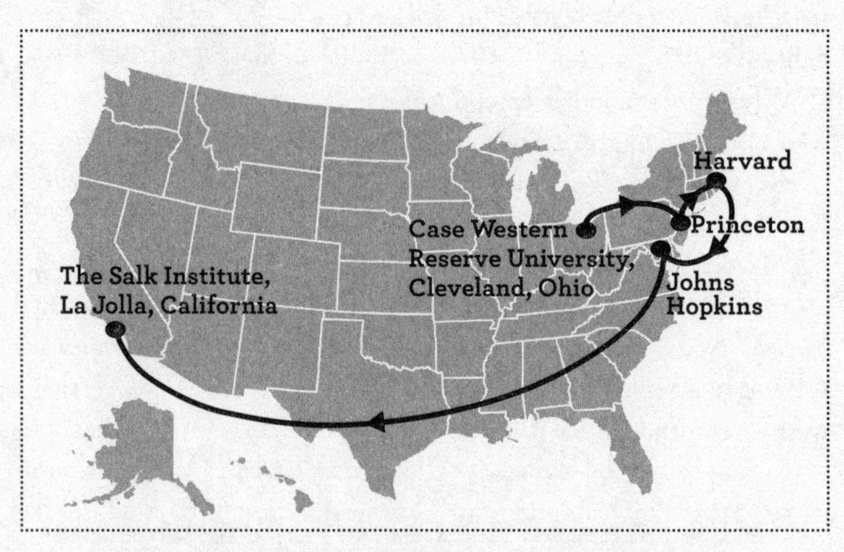

Terry's studies and work would take him to some of the leading research institutions in the United States. Today, conferences and collaborations with fellow researchers bring him around the world.

Princeton had a process in which, in one week of grueling general exams, graduate students demonstrated all the physics they'd learned from classical mechanics, quantum physics, electricity and magnetism, thermodynamics, statistical mechanics, condensed matter physics, and particle physics all the way through general relativity.

These kinds of qualifying exams are common in doctoral programs all around the world. What is uncommon is that the world-class Princeton physics professors, happily engaged with their super-smart students, began to create exams that were more and more challenging each year.[3] Each professor devised questions that delved ever more deeply into the most thought-provoking intricacies of their specific areas. When the questions were all put into one big qualifying exam, that exam started to become not just demanding, but off-the-scale, next-to-impossible difficult. Great students began washing out of Princeton's physics program.

Finally, someone had the bright idea to give the full exam to the physics professors themselves. Some very smart people flunked it. The questions were scaled back.

But before then, Terry took the incredibly difficult test. He aced it.

Stepping Back to Look at a Career Big Picture

Terry's master's degree focused on general relativity. String theory, he found, was becoming the only game in town for theoretical particle physicists, and it was becoming more and more esoteric. Experimental studies needed big explosions in space or massive accelerators to make even the tiniest advance. Accelerators were needing ever more energy. At some point, physicists were beginning to realize, it would take the annual budget of the United States to build a big enough particle accelerator to make real progress. Meanwhile, cosmology had similar issues—it needed wildly expensive satellites and giant interferometers.

These issues were initially storm clouds on the horizon—isolated from the day-to-day toil of pulling physical ideas, like taffy, from the math. Terry enjoyed the intense rush of discovery—the *click* feeling as new findings and theories sprang to life in his mind, sometimes making fresh connections even the original theorists hadn't seen.

But once past the doctoral exams, Terry wasn't all work. He was a gregarious guy—had fun with friends, went to movies and out to dinner. He also had a charismatic girlfriend who was smart, vivacious, and beautiful. The reality was that Terry was the perfect catch for any family with intellectual aspirations. It was hard to be more academically serious than to be a student of relativity at Princeton, mentored by one of the greatest advisers in the world. Career-wise, Terry was set to soar.

But then the storm clouds came in closer, making Terry question his commitment to physics. What kind of impact could he make in foundational research on relativity when throughout his career, he would be hearing, "We just can't afford to build the device you need"? He'd invested so much into his first love, physics, that it was hard to imagine shifting paths. But he just couldn't find a way to push away the thought: *Should I be looking elsewhere for my career?* Was there a place where he wouldn't always be battling the constraints of enormous costs? It felt almost sinful to ponder that in the bastion of the physics group at Princeton, where advancing knowledge of relativity was a sort of research Holy Grail.

Despite—or perhaps because of—his passion for physics, Terry was interested in everything. He had friends in biology, so he decided to take a course from noted neurobiologist Mark Konishi on neuroethology.[4] This is a way of applying ideas from physics to the study of natural behavior, such as how owls use sound to work out where their prey is and how baby birds learn to differentiate their own species' song from the hundreds of other songs they hear.

Intriguing ideas in biology started coming to Terry from different angles. A lecture by visiting professor Chuck Stevens, from Yale, revealed that synapses, the connectors that allow neurons to talk with one another, are unreliable. "How," Terry wondered, "could the brain compute with such unreliable parts?"[5] He went to a meeting of the Society for Neuroscience and was surprised at the size and enthusiasm of the crowd.

There were, Terry had begun to realize, two very different universes. There was life *outside* the brain—which included the billions-of-light-years-broad universe as well as the femto-level tininess within atoms. This splendid collection of macro and micro was all within the realm of physics.

But there was also the universe *inside* the brain: the unknown, seemingly mystical home of our thoughts and feelings and of consciousness itself. People had just started using a new term for studying these subjects: "neuroscience." However, neuroscience, at least in the late 1970s, didn't have the gravitas of relativity studies—instead, it was still a toddler in the firmament of science. Careers in neuroscience didn't even seem possible. Biology itself seemed a poor relation to the vaunted study of physics.

Meanwhile, Terry's girlfriend's parents were aghast. Here Terry was, a top physics student, slumming around with an interest in *biology*? As far as they were concerned, Terry was an academic playboy—not serious about establishing a world-class career.

After a few tense weeks, his girlfriend broke up with him.

It was an emotional gut punch, but the experience made Terry take a second look at the world and his place in it. He began hanging out at the labs of Charles Gross and Alan Gelperin, two professors of neurobiology at Princeton. Instead of working with John Wheeler in relativity, Terry would ultimately have a doctoral adviser who had himself transitioned from physics to neuroscience: the dazzling scholar John Hopfield. Hopfield had done groundbreaking work in the late 1950s with polaritons, a sort of "almost" particle where electrons couple with surrounding solids. Among other breakthroughs, Hopfield eventually developed the famed "Hopfield net," which has led to a better understanding of the neurocircuitry that underlies memory. Terry's transition from relativity to neuroscience happened over several years. During this time he led a double life, going to biology classes during the day and writing his physics thesis at night. Encouragement from John Hopfield made all the difference. Terry went on to publish a series of papers on models of neural networks that were inspired by the groundbreaking, ultimately Nobel Prize–winning work of David Hubel and Torsten Wiesel on the visual cortex. Terry's own papers became his doctoral thesis: "A Stochastic Model of Nonlinearly Interacting Neurons."

Watch Out for Ruts

In science, researchers frequently spend years mastering a technique that allows them to grapple with a particular set of problems. The technique might center on, say, a type of imaging or a way of statistically analyzing data. An entire career frequently emerges from variations on this technique.

"The process of variations on the same theme is not unique to science," Terry observes. "You get a skill. You use that skill over and over again. But after a while, you get stuck in a rut. You're bored. Or the field changes and you begin to realize you need new skills. However, you may not have gotten the skills early that you need for a new direction you'd like to pursue. In science, it's especially difficult in that, after a decade of effort, you may have a PhD in one narrow field. But you're still an amateur in other fields."

Terry's doctoral adviser, John Hopfield, had shown firsthand that it was possible to make a career transition between physics and biology that could pay off. And Terry was convinced that there were ways to apply the mathematical modeling tools of physics to better understand biology and, particularly, neurons. But he also knew that he still didn't have the in-depth biology he needed to be an expert in neuroscience. As his mentor Alan Gelperin would later say, Terry needed to get "neurons under his fingernails."

But then, even if Terry was able to get the background he needed in neuroscience, how could he break into the field? Despite growing interest in the field from researchers, there weren't many established neuroscience departments around the country. Finding a job could be tough.

Networking to Find the Right Place

Harvard was pretty much *the* place to be if you wanted to be on the inside of research in neurobiology. However, Terry was doing his doctorate in physics at Princeton, hundreds of miles away. He was in the wrong place and the wrong field, even if he was quietly twisting the edges of acceptable physics research with his interests in neurons.

As it happened, in the summer of 1978, there was a course on neurobiology at Woods Hole, a research center on a corner of Cape Cod. Terry signed up. He had heard rumors that Woods Hole was a casual place, so he showed up with standard Princeton attire—a white shirt and suit coat. To meet the "casual" standards, he left off his tie.

This immediately made Terry the butt of a series of good-natured jokes from both fellow attendees and summer school faculty. Neurobiologist Story Landis, the woman who would go on to become the director of the National Institute of Neurological Disorders and Stroke at the National Institutes of Health, bought Terry his first pair of jeans. But Landis did more for Terry than simply expand his wardrobe. With her help, and that of others at the summer course, Terry began his magical summer of 1978, throwing himself headfirst into a new discipline.

The course was difficult—Terry had never worked so hard in his life. At the same time, it was thrilling—the group was taught by some of the best neuroscientists in the world. The class ran from June through August. In September, however, Terry was still in Woods Hole, finishing a project on skate electroreceptors that would eventually lead to his first paper in biology.

One day, while he was sitting in the lab at Woods Hole, the phone rang and Terry picked it up. It was Harvard neurobiologist Steve Kuffler. Would Terry like to come to Harvard to do his postdoc studies with Kuffler? It was like getting a call from Saint Peter himself—Kuffler was often referred to as the "father of modern neurobiology." In fact, Kuffler would have been a shoe-in for the Nobel Prize, save for his unfortunate later career move of dying too early. (Nobel Prizes are given only to the living.)

Kuffler's call was a sign—Terry was moving up to the big leagues. But it wasn't quite as easy as all that. There was some rapid shuffling as Terry wrapped up his doctoral dissertation. Then, off he went to join Kuffler at Harvard.

 **Now You Try!**

Competence Is Key

Terry went to Harvard because he was in an extremely specialized discipline, and that was the best place to gain the expertise he needed for the research he wanted to do. ***Knowing your subject matter is key.*** For a restaurateur, for example, knowing your subject matter doesn't mean going to Harvard. It means understanding every aspect of restaurant operation because you've done it yourself, from busboy to manager.

Think about an area you're already expert in, or that you *want* to become expert in. Under the title "Competence Is Key," jot down some key ideas about what you have done in the past, or what you need to do in the future, to truly master the subject matter that holds your future.

The Value of Selective Ignorance

Terry's strong technical background created a surprising disadvantage for him. He knew he could easily be made into a technician rather than a real biologist. (*Hey, this new guy Terry seems to know his way around a computer—let's get him to write programs.*) Because of this, Terry vowed not to touch computers for the three years of his postdoctoral studies. Instead, he did nothing but live and breathe neurobiology.

Terry's painstaking focus on the core of his new discipline paid off. Even if he wasn't the top postdoc emerging from Harvard in the field of neuroscience, he was still outstanding, like everyone else from the program. But Terry was different from the other students. Hidden beneath his new array of neuroscientific tools was a comprehensive knowledge of physics, with its rich ways of modeling the world. Even Terry didn't yet realize the full power of his mental arsenal.

Harvard prepared its postdocs to know how to speak, not only to experts, but also to others who don't have a background in the field. At Harvard, Terry learned how to build intriguing story hooks for everyone from beginners to experts.

This relates to how I first met Terry, when I gave a presentation on my scientific research—with a few embedded story hooks—for the National Academy of Sciences. The event took place as part of the Sackler Colloquia at the Beckman Center in Irvine, California, where I felt like a cub in a lion's pride of world-class researchers. Terry was serving as moderator, and he put me instantly at ease. We became friends over shared wonderment about how people learn and change.

I told a former boss I wanted to learn how to do everything at the company. Her response? "Don't do that. You need to cultivate 'selective ignorance,' because if you know how to do everything then you become everyone's bitch." Turned out to be good advice. Cultivating "selective ignorance" has often prevented me from getting drafted into projects I had no interest in participating in or time for.

—Brian Brookshire,
online marketing specialist at Brookshire Enterprises

Key Mindshift

Selective Ignorance

You only have so much cognitive energy. Be selective about what you choose to become expert in—you don't want to be typecast as an expert in areas you do not want to spend time on.

Keeping an Open Mind

A year and a half after my National Academy of Sciences presentation, my hubby and I spent a sunny July day at the glide port by the Salk Institute near San Diego with Terry and his brilliant, medical doctor-researcher wife, Beatrice Golomb. (I ended up peeling like a boiled tomato.) Terry reflected on creativity and career change as we watched hang gliders and paragliders soaring from the cliff's edge out into the skies above the ocean nearly four hundred feet below.

"There is a disadvantage to being well integrated into your discipline," Terry observed. "Different disciplines have different cultures. The more comfortable you are with one culture, the tougher it can be to shift to another."

Neuroscience, with its exciting potential for breakthroughs, has become a target for many as they've changed careers—Terry has had a front-row seat in watching people transform themselves. "Career shifting can be like starting a new relationship," he notes. "It's a process that can take years—but it's exciting and rejuvenating. For example, even if you stay only in the field of medicine, rejuvenation can come through moving from one specialty to another."

The rejuvenation can also lead to career-enhancing breakthroughs. After all, new insights often come the first time you're learning something new. Once you've absorbed the material, it closes your mind off to looking at the material in fresh ways. "It's *not* your chronological age," Terry continues. "It's how many years you're in your field." But being on the forefront of scientific research isn't easy. As Terry says, "You can always tell the pioneers by the arrows in their backs."

Of course, part of grasping new insights is also being open to what the facts actually tell you—instead of what you *want* the facts to tell you or what everyone agrees the facts say. Beatrice was the first researcher to publish findings revealing that cholesterol-lowering statins, which sometimes prolong life, can also pose problems such as muscle pain and memory issues.[6] But getting those facts into publication wasn't easy for her—journal reviewers felt uncomfortable approving findings that ran counter to the expectation that statins were a largely unalloyed good.

The Importance of Humility

Terry has reflected deeply over the years on the fields he has been associated with. Physics, Terry has observed, is a field imbued with hubris—the academic equivalent of Wall Street "Masters of the Universe." (Even so, I have to admit, one of the reasons I like Terry is for his lack of hubris. Like many highly intelligent people, he leaps to conclusions that are sometimes wrong. But unlike many, he is quick to correct himself and change course once he's detected his error. Also he's one of those rare people who isn't wedded to an idea just because he thought of it. In this way, he's surprisingly different from many academics.)

Physicists have a tendency to believe that physics is the most difficult subject to master and that physicists are the smartest researchers. The reality, of course, is that physics is a field that does have a lot of intelligent people. And this makes it all the more interesting to observe their more spectacular goof-ups.

One distinguished theoretical particle physicist at Caltech who Terry knew decided to go into neuroscience. Despite his lack of knowledge about the field, this top-flight researcher built a lab, hired a talented postdoc, and began ordering him around. It all blew up—the lab fell into disarray and was dismantled. The reason? You can't assume that just because you've mastered field A that you can master field B. It's easy to think a research idea is going to be groundbreaking if you don't know enough about the field to understand that your idea isn't feasible or has already been explored.

Experimental particle physicist Jerry Pine typified the other way to go into neuroscience. Jerry was a full professor at Caltech, at the pinnacle of his career, when he decided to make the shift. He took the Woods Hole neurobiology course with Terry (arriving in jeans, Terry noted sheepishly). Jerry and Terry were, in fact, the only physicists in the course—all the others were biologists. Afterward, Jerry packed his family up and took off for three years as a lowly postdoc at Washington University in St. Louis. Eventually, he went on to build electronic chips that neurons can grow on, giving researchers a better understanding of how neurons interact with one another in groups.

Learning a second discipline, as both Terry and Jerry did, takes time—maybe a lot of time if the second discipline is very different from the first. You will need to find a place where you can learn the ropes and where other people will help you. There are likely to be setbacks in the learning process. In fact, at first, it may feel like taking two steps back for each step forward, but if you're lucky, you can marry some of your old skills with new ones.

To avoid falling into a career rut, it's important to be open to change, like Jerry Pine. Humility is important during the learning process, along with persistence. These traits allow you to better place yourself into a new context—and as we'll touch on next, it is *context* that allows you to make the shift.

Key Mindshift

Learning a New Discipline Takes Time

If you are learning something new and difficult, explore boot camp–like experiences to make new connections and immerse yourself in the new ideas. No matter how smart you may be, give yourself the time you need to truly learn the discipline.

Context Is King

The circumstances in which we perceive something—its *context*—have an enormous effect on how we react to it. Your actions would be different if, a few feet away from you, you saw a poisonous snake poised to strike in a glass cage as opposed to skittering toward you on a table.[7] We constantly take in all sorts of cues from the environment, and also from our own thoughts and feelings.

This, in fact, is why the placebo effect can be so powerful. Our conscious thoughts, which are formed in the prefrontal cortex, can spur physical changes throughout the body. For example, if a nurse tells us that a procedure is going to hurt, levels of stress hormones rise within seconds. This can make the experience more painful through the nocebo effect, which activates the pain-boosting "CCKergic" systems.[8] Similarly, if we ***believe*** that a particular substance will reduce pain—even if it turns out to be sugar water or saline solution—that belief can activate the body's natural opioid systems and reduce pain.[9] The placebo effect is so powerful that once it's been in place for just a few days, the effect persists—even after people are told that they aren't getting the real drug.[10]

It isn't just pain-related systems that can change with differing perceptions. For example, if we believe a first type of milkshake is more filling than a second type, the first milkshake will do more to reduce levels of the pro-hunger hormone ghrelin.[11] Swallowing a strange-tasting drink that has an immunosuppressive drug in it can eventually produce immune suppression from the taste alone.[12] And anxiety-reducing drugs that reduce unpleasant reactions to fearful and threatening pictures can later produce the same effect when they're replaced by a placebo.[13]

All in all, your expectations about what will happen, and the underlying context, can powerfully shape your mind and body's reaction, both for good and for ill. It was what underpinned Claudia Meadows' escape from depression. It is what allows for the dramatic success of cognitive behavioral therapy.

And it can also underpin your own success in whatever you are trying to learn and become.

Immersion

Terry and Beatrice's friend Francis Crick was keenly aware of the importance of context in changing careers. As the codiscoverer of DNA, the secret code of life, Crick was a towering figure in science. Crick had already reinvented himself once in his early thirties—a career change that underpinned his Nobel Prize–winning breakthrough. He'd been a promising physics student at University College in London until a World War II German bomb fell through the roof of his laboratory and destroyed his equipment.

Delayed for years during the war by his work designing mines that could avoid German minesweepers, Crick finally took up the study of biology at the advanced (at least in the science world) age of thirty-one. For Crick, as for Terry, the transition from physics to biology was difficult—Crick had said it was "almost as if one had to be born again."[14] Despite the difficulty of switching from physics' "elegance and deep simplicity" to the complexity of biology's evolving chemical mechanisms, Crick felt, oddly enough, that his original training in physics had given him something of great value—that double-edged sword of hubris. Proud Masters of the Universe—his colleagues in physics—had made enormous breakthroughs. If they did it in physics, why couldn't he do it as well in biology?

And thus unfolded the career change from physics to biology that underpinned Crick's vitally important role in the discovery of the structure of DNA. However, one major, Nobel Prize–winning career change wasn't enough for Crick. As he neared sixty, an age when many people begin to slow, Crick developed an interest in one of science's toughest problems: the origins and workings of human consciousness. Unlike many at the time, Crick had a gut feeling that the underlying neuroanatomy was key. To begin to understand consciousness, then, he needed to delve into neuroscience.

Crick's challenge, however, was that he was too good at what he'd already mastered. The profoundly important discovery of DNA and its accompanying Nobel Prize meant that he found himself locked with golden handcuffs to the throne of laboratory molecular biology, with his world-class research digs at Cambridge forming a sort of scientific prison.

To break out, Crick decided to move from England to San Diego, to the Salk Institute. *He shifted his context.* In this new, sunny environment, his day-to-day interactions weren't with molecular biologists—they were with neuroscientists. "He'd have dialogues that would go on for days," Terry recalls. "He'd ask people to come and train him through discussions."

Crick immersed himself in his new discipline. Though he wasn't able to solve the problem of consciousness (a tough nut to crack), he played an important role in moving consciousness studies onto a solid research foundation. He was busy editing his latest article on neurobiology just days before his death at age eighty-eight.

Changing and learning something new, as Terry did when he was younger and as Crick did in later life, is absolutely possible. Research is giving us clues about how we can improve our abilities to learn and change—even as we age.

Never Too Old to Learn and Change

Surprisingly often, we feel guilty about changing our careers or learning something new. When we are in our twenties, we think, "I could have been a first-rate guitarist if I'd just started when I was a kid!" When we reach age sixty, we look wistfully back at the more open possibilities of our thirties. We forget that when we were in our thirties, our options often seemed equally limited. Even college freshmen look with envy on other students who began studying French, physics, or philosophy in high school. No matter what our age, we often feel too old to learn something new.

Often, it's hard to realize that the path not taken always seems alluring—and to see that there are benefits to the path you ***have*** taken. Retraining your brain to master something new as an adult can have profound benefits—not only for you, but for those around you and for society as a whole. These benefits are so valuable that you may be surprised to learn that even the world's most accomplished people actively seek out career change. Some even plan for regular changes in their careers ahead of time. Stephen Hicks, a professor of philosophy at Rockford University in Illinois, observes:

> When I was a grad student and becoming serious about a career in philosophy, I was impressed by a profile I read about Subrahmanyan Chandrasekhar, the physicist. His strategy was to read and think intensively for several years in one area of physics and then write several papers and an overarching book to integrate the ideas. Then he would shift to another, often quite different, area in physics and do the same thing. Over the decades, he avoided rut thinking and was able to make creative contributions to many areas.
>
> Since philosophy is such a sprawling discipline, and since

one feature that attracted me to philosophy is its fundamentality to many different intellectual areas, I resolved to follow Chandrasekhar's strategy. Since finishing grad school, my career has been in six-year units—four years of reading, thinking, and writing shorter articles in an area and then two years to finish a book. Then I jump to another distinct area.

The six-year pattern was not mechanically scheduled but emerged organically. And while I've worked in several distinct areas now, there are nonetheless connections between them, so I expect and plan that by the time I am done, I will have completed work that integrates to form an overall philosophy.

Setting Up a Context That Works for Big Learning Changes in Your Life

Shifting your mind-set about your capabilities often isn't easy. Those around you can sometimes conspire to keep you where you are instead of where you want to go. There are different ways to deal with this challenge.

- **Leave:** If the situation is toxic, "damn the torpedoes" and remove yourself from it. This was the approach Zach Caceres took in dropping out of a toxic high school.
- **Double life:** Live a double life for a while, simulating the old lifestyle and interests while developing new interests on the side. This approach worked for Graham Keir and Terry Sejnowski—it prevented each from getting into a situation where others could continually argue against a switch.
- **Contrarian:** Take pride in being a contrarian. The more others say you will fail, the more it can bolster your internal resolve. This worked for

Adam Khoo, who set up interim goals—like getting into a well-known junior college—that proved to himself as well as to others that he could achieve his aims. Keep in mind, however, that it's important to choose reachable and doable interim goals and checkpoints to assess your progress. For example, if you try as hard as you can but repeatedly get a very low score on the MCAT, it may be time to reassess your dream of medical school.

If you are lucky, those around you will support you in your attempts to change. Rejoice and use the opportunity to go as deep as you can into the learning experience. This is what physicist Jerry Pine did when he moved himself and his family from Caltech to St. Louis and went from full professor to lowly postdoc to retool his career.

Don't play up mental roadblocks that will keep you from a newly discovered passion. But don't minimize important considerations either—like whether you've got at least the basics of what it might take to succeed. You don't want to be like a cluelessly bad karaoke singer warbling haplessly into the night.

You Can Teach an Old Dog New Tricks

Terry has developed powerful computer modeling techniques that support our understanding of complex phenomena like memory, thought, and feeling. This means he has an astonishingly broad understanding of many different facets of neuroscientific research.

"As you shift into later stages of your life, learning something new can become slower and more difficult," Terry says. "But you can still do it—the brain is still plastic. What's especially interesting is that there are breakthroughs on the horizon that can prevent cognitive decline with age."

As we age, we tend to lose synapses and even neurons, like water trickling out of a dam. But it's not an entirely losing proposition—not, that is, unless we let it. Exercise, learning, and exposure to new environments can help create and nurture new neurons and synapses. Activities like these serve as a sort of cognitive rain that replenishes the water behind the neural dam. This builds up what's known as a "synaptic reserve," which is especially important as you grow older, to help balance the neurons and synaptic connections that are being lost.

I asked Terry which researchers he thinks are extraordinary in uncovering ways we can keep improving our brains as we age. "Daphne Bavelier," said Terry, without hesitation.

Bavelier, a cognitive neuroscientist at the University of Geneva in Switzerland, studies video games—shoot-'em-up, action-style video games. What she has found has upended stereotypes about how bad video games are for you—and has given insight into future therapies to keep the brain in top shape even as we get into our "golden years."[15]

Conventional wisdom is that too much screen time with video games worsens eyesight. To Bavelier's amazement, when she quantified the eyesight of action video game players, they had *better* vision than average. That vision was better in two subtle but important ways—action gamers were better able to pick out minute detail within a field of clutter. They could also pick out more levels of gray.

This difference might seem shrug-off-able, but translated into real-world terms, this means action gamers can drive better in foggy conditions, and as they get older, they can read the fine print on prescription bottles without using magnifiers. In other words, video gaming allows us to improve in some of the very areas that can cause danger and difficulty for people as they age.

But Bavelier and her colleagues have discovered even more.

Many people believe that video games lead to distractibility and problems with attention. But when it comes to action video games, the opposite occurs. In Bavelier and her colleagues' studies of action gamers, they have observed that key "focusing" areas of the brain become much more efficient. Action gamers can also switch their attention quickly with only a small mental cost. In essence, gamers concentrate better. They can, for example, more easily flick their focus from the road in front of them to the dog darting in from the side.

Basically, it seems that action-style video games allow us to improve on many of the areas that begin to falter for us as we age. As Bavelier notes, "Complex training environments such as action video game play may actually foster brain plasticity and learning."[16] Action video games not only seem to help us see better, focus better, and even learn better, but the effects last a long time, showing up even many months later. (Incidentally, if you want to improve your ability to do spatial rotation—an important skill in both art and engineering—*Tetris* is for you.)

As far as neural improvement, the brutal-but-engrossing *Medal of Honor* beats *The Sims* hands-down. This may be because in *The Sims*, you do not have much demand on attentional control. In *Medal of Honor*, however, your attention changes around to different areas on the screen, from very spread out as you monitor your environment for new enemies to very focused as you need to aim accurately. *Medal of Honor* also gets you viscerally involved in play, with background music and lots of unexpected changes and looming motions that attract attention at multiple neural levels below conscious awareness.[17] This kind of attention-grabbing may lie at the heart of plastic changes.

So why don't we already have great video games that are specifically designed to correct age-related declines? Bavelier equates it to coming up with tasty chocolate (the video game) that's combined with healthy broccoli (the cognitive enhancement).[18] Combining chocolate with broccoli in a form people want to consume is not easy—not even for a master chef. But brain scientists working together with artists and the entertainment industry are making progress.

Of course, common sense is also important when it comes to video games—researchers agree that excessive bingeing isn't healthy. But fortunately, bingeing isn't needed—positive effects were only found in those playing for short periods, thirty minutes or so per day, regularly over a period of a few months.

Retaining the ability to learn and change as an adult is a multifaceted challenge. It goes beyond just interacting with a video game, textbook, fellow students, or a teacher. Physical exercise, as we've mentioned, is vitally important. Pharmaceutical drugs such as Ritalin and Adderall can sometimes also enhance our ability to learn. But they're loaded with unwanted side effects—the equivalent of throwing a bucket of lead-based paint at a tiny speck of dirt on your living room wall. Good nutrition can be a boon as well, although at some point it's hard to make more improvements in cognitive function even as you've begun to step into the choppy waters of contentious nutritional theories.

But action-style video games are gaining the attention of leading researchers because of their extraordinarily broad impact—they provide easy ways to see how sights, sounds, actions, and activities can produce changes in key learning processes like attentional allocation, resistance to distraction, working memory, and task switching. We can already see how certain types of gamers use their brain more efficiently—needing fewer neural resources for demanding tasks. Gamers are also better at suppressing irrelevant information.

Building in part on Daphne Bavelier's decade-plus body of research, UC San Francisco researcher Adam Gazzaley—another of Terry's recommended researchers—also focuses on video games. Gazzaley, who is both a neuroscientist and neurologist, points out that video games are among the most powerful forms of media. They're both interactive (and isn't that what teachers are striving for?) and fun. Gazzaley is attempting to create a more potent mix of gaming and therapy—and he's succeeding. *Nature*, one of the world's most prestigious scientific research journals, put Gazzaley's research on its cover under the title "Game Changer."[19]

Gazzaley's approach to developing new therapies is indeed a game changer. His *Neuroracer* is a deceptively simple game in which you drive a speeding race car along a roadway where signs randomly pop up, forcing you to react.[20] Gazzaley found that older subjects who played just one hour of *Neuroracer* a day, three days a week for a month—twelve hours of total play—experienced a strong and lasting improvement in concentration. The game is going through an FDA pathway, and Gazzaley hopes to see it become the world's first prescribed video game.

It turns out that your ability to focus your attention, hold something in working memory, and keep other thoughts from intruding arises from what's called the "midline frontal theta." This is a burst of electrical waves that appear toward the front of your brain when you are engaging your attention.[21] But it isn't just the front of the brain that's important when you're trying to concentrate. The front of the brain also needs to be able to send signals and communicate with the back of the brain. Technically, this relates to the "long-range theta coherence." As you grow older, the power and coherence of these interconnecting brain waves can weaken. Decline in the midline frontal theta and the long-range theta coherence is one reason why older people can find themselves standing in the kitchen wondering what they came in for. It's also why their driving reactions slow.

Neuroracer gives people the ability to practice and improve their ability to focus. It's fun to boot. What's most important in all of this, though, is that we can see *why* the improvements happen. *It's due to the changes in theta rhythms.* As Gazzaley's research has shown, with a gam-

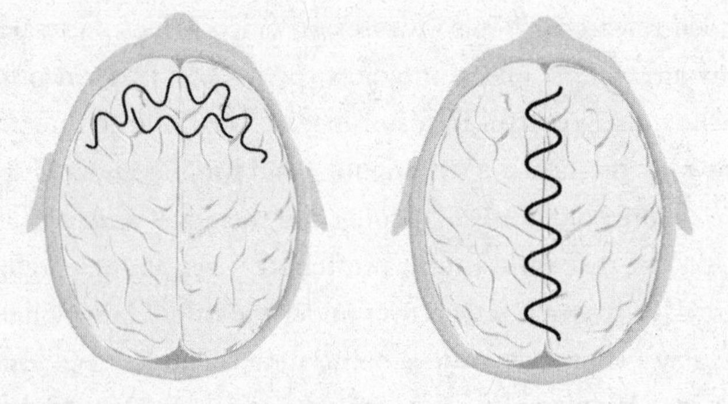

On the left, we can see lines that represent the "midline frontal theta waves" that appear toward the front of the brain when you are concentrating. On the right is a wave representing the theta wave communication between the front and the back of the brain. Both of these types of wave activities can decline with age—but can be ramped back up through the powerful effects of playing video games.

ing approach, sixty-year-olds can learn to outperform twenty-year-olds! *Neuroracer* appears to pinpoint the neural markers that serve as pivots for many of the most important cognitive skills, like working memory and vigilance, and enhances them. That means these skills are also improved, even though the game doesn't specifically target them.

We're beginning to learn the elements of a cognitively winning game system. Art, music, and story can create the kind of immersion and engagement that produces ideal conditions for neural plasticity. A good game, in other words, creates a sort of neural retooling kit to mold and shape cognition. There's also evidence that video games may combat the ill effects of ADHD, depression, dementia, and autism.

Gazzaley's goal is real-time feedback. He's working to create a system that pinpoints weaknesses in neural processing and uses the information to challenge the player. Weakened neural signals could be improved in an easy, entertaining way. "What would it be like to enter your own brain," Gazzaley asks, "with your challenge being to improve the neural processing you're seeing? You could learn to control how your brain is processing information."[22]

On another front, neuroscientists Mike Merzenich and Paula Tallal developed computer-based exercises to allow people with dyslexia to more easily distinguish certain sounds. This in turn can make dramatic improvements in reading ability. The results of this innovative research were published in *Science*—and resulted in a deluge of more than forty thousand phone calls from parents desperate to improve their challenged children's ability to learn.[23]

Merzenich recently won the Kavli Prize, which is the equivalent of the Nobel Prize for neuroscience—he's also a member of the National Academy of Sciences as well as the National Academy of Medicine. In other words, Merzenich is the real deal as a highly respected scientist. Building on the success with brain-training for dyslexia, Merzenich has created a company, Posit Science Corporation, that focuses on improving cognitive performance. BrainHQ, the company's flagship product, isn't meant to make you into a genius—instead, it is structured to allow you to reach and maintain your cognitive peak by giving you exercises to speed neural processing, strengthen attention, and improve working memory. Reputable studies show that BrainHQ does appear to make a difference, whether it's in helping you remember people's faces, being more responsive while driving, or keeping up with rapid-fire conversations.[24]

There are hundreds of brain-training programs available online—most have less-than-compelling evidence that they work. But top scientists like Bavelier, Gazzaley, and Merzenich are leading the way in showing that "mental therapies" really can make a difference.

Building a Cognitive Reserve

We know that some fourteen hundred new neurons are born every day in the hippocampus. There's only modest decline in this neural birth rate as you age.[25] But unless the brain continues to encounter novel experiences, many of these new neurons will die off before they mature and hook into the larger neural network, rather like vines that languish and die for want of a trellis.

In adults, new "granule" neurons allow us to distinguish between similar experiences and store them as distinct memories. These newer cells are different from older cells, which carry patterns that associate similar memories to one another.[26] New neurons are especially valuable when it comes to avoiding the rekindling of older, sometimes more traumatic memories.[27] All this means that, for new learning as well as mental health, it's important to help new neurons to be born, survive, and thrive. This is why neurogenesis has become a hot area in treating depression and various anxiety disorders.[28]

Of course, as mentioned earlier, *exercise* is one of the most powerful "medications" we know of that produces new neurons. It's as if exercise scatters seeds that become neural sprouts. *Learning*, on the other hand, is like water and fertilizer that encourage the growth of those neural sprouts.

The younger you are, the more likely it is that anything you experience is novel. As you age, it gets easier to fall into a rut. Even when you tell yourself you're learning something new, it's often just a slight riff on something you already know. Learning that makes an impact on the brain often means going just a bit beyond your comfort zone.

Larry Katz, a neuroscientist at Duke, suggested that a useful way to allow the new neurons to survive, thrive, and make new connections is to do something new and different every day.[29] This automatically presents your brain with novel experiences. These novel experiences can be as simple as using your left hand to brush your teeth if you're right-handed. Or just sitting in a different chair at the dinner table. This is also why travel can be so invigorating. It keeps the brain tuned up, especially if you do your best to immerse yourself in the new culture and surroundings. Learning a foreign language when you are older may also be especially worthwhile, because the areas of the brain positively affected by language learning include many areas that are negatively affected by aging.[30]

In brain terms, if you don't use it, you can lose it—no matter how innate and natural your gifts might seem. Widely admired orator Robert Sobukwe, who spoke so eloquently for the cause of liberation of black South Africans from the rule of apartheid, was subjected to six years of solitary confinement on remote Robben Island. He could communicate with other prisoners only through furtive hand signals. During this terrible time, Sobukwe could feel his powers of speech slipping away.[31] Those who have endured the winters at remote Antarctic stations with few opportunities to speak with others have experienced a similar sensation—and discovered themselves stumbling through simple conversations upon their return to civilization.[32]

A variety of hobbies keeps us mentally tuned up—especially when those activities are combined with exercise. If you knit, sew, quilt, do plumbing or carpentry, play games, use your computer, or read, for example, research shows you're more likely to have stronger cognitive ability as you age.[33] These findings make sense—for example, measuring and cutting for either quilting or carpentry clearly help maintains your spatial abilities.[34] On a side note, one recent controlled study found that those who read books for three and a half hours or more a week were 23 percent less likely to have died over the twelve-year study period.[35] It's definitely books that had the effect—magazine and newspaper readers didn't fare so well. (Yay to you and your longer life for reading this book!)

One intriguing study of more than sixteen thousand participants in rural China showed that the probability of Alzheimer's disease was distinctly correlated with a person's level of education.[36] This also makes sense. The more the intellectual stimulation, the lower the Alzheimer's risk. Yes, it's just a correlational study—we don't know for sure that the intellectual stimulation is truly the *cause* of the lower Alzheimer's risk. But we do know that more education produces more synapses, and the more synapses, the bigger your cognitive reserve. In any case, education isn't just something to get a dose of when you're younger. Studies have shown that the more active a "learning lifestyle" you have as an older person, the lower your risk of Alzheimer's.[37] What you learn as a mature adult, or as an older person, continues to build and maintain your cognitive reserve.

Key Mindshift

Comfort Zones and Synaptic Reserve

Day-to-day activities as simple as talking, knitting, or shooting hoops keep us in mental as well as physical shape by retaining abilities we already have. But when we go a bit beyond our familiar comfort zone by learning something that challenges us, it helps build a synaptic reserve. This reserve is increasingly important as you age.

Learning and Changing at Any Age

Just as Terry surmised, there have been few major advances in his former area of physics in the last decades, and exorbitant equipment costs have played a role. Many of Terry's friends who did go into particle physics ended up looking for jobs in other fields.[38] The allure of a seemingly hot area, the "sheeple" mentality, and lack of knowledge about limited opportunities are phenomena that arise in many careers and vocations. In any academic discipline there can be a lemming mentality, with professors encouraging students to major in their particular specialty even if the job prospects are dim and tuition costs are stratospheric. Students look to one another, thinking, *Hey, professors wouldn't be so encouraging if it was a bad idea.*

Despite the fact that he was at the pinnacle of prestigious physics studies at Princeton, Terry Sejnowski used common sense to step back, assess, reevaluate, and make a reasoned bet that a career change was in order. This was despite the fact that such a change would be difficult, and that at the time, few were making it. In the final analysis, the benefit from Terry's willingness to risk his career and head where he felt he could make the greatest scientific and social impact has been enormous.

Quantifying how neurons speak to one another means that we can better conceive of our essence as human beings—how we form memories, why we can smell a rose, how we hit a baseball, and why we dream. Thanks to Terry and his colleagues' work, we now better understand how the brain works, how to tease more useful data from research analyses, and how to make predictions that can maximize chances for research breakthroughs. The algorithms and tools Terry has developed have helped researchers worldwide.

Among many other areas of research, Terry Sejnowski has worked to reveal the importance of exercise in cognition and learning. He makes a point of incorporating exercise into his daily routines, wherever he might be. Here he enjoys a break at the Waterton Lakes National Park in Alberta, Canada.

But what happens when you're looking for a career, and unlike Terry, your promising dreams and opportunities are crushed early on?

In our next chapter, we'll meet Princess Allotey, who will show how youthful resilience, and willingness to take advantage of unexpected opportunities, can make all the difference.

A Career in Neuroscience?

To have a career spearheading scientific research of any kind, you need to put in the time, effort, and money to get a doctorate. This is all before you could even begin to *think* of the intensely competitive process of applying for a tenure-track job at a university. It's typical these days to have hundreds of applicants for a single position.

Neuroscience is now so popular that there's reason to be cautious when considering the field. But as neuroscientist Alan Gelperin points out, competition has always been heated in most academic fields, including biology, physics, engineering, and of course neuroscience.

Princeton neuroscientist Alan Gelperin has been working in neuroscience for more than fifty years. His valuable insights on competition apply to many fields.

Alan asks, "Which is your favorite trendy area that's currently sucking people in? Developmental biology? Molecular biology? The powerful gene-editing technology CRISPR? Want to modify some genes? Get a cookbook—you can order it off the web. Get some eggs, and God knows, you can make a frog that can talk."

A career in scientific research often includes risk-taking. One of the biggest risks is that someone else publishes the results you were just planning to publish. Alan suggests taking a "sufficiently unique combination" approach, where your idea and your skills combine to make it reasonably unlikely that, by the time you've started to make progress, another researcher will suddenly scoop you.

Even so, there are famous instances of people who followed that reasonably unique approach only to pick up a journal after five years of research and—*oh my, what a great paper!*—someone else just published the findings you were about to put out. Of first order importance, then, is figuring out something that's unlikely to appear in a journal in the time period it's going to take to make some progress.

"Going into a new area, you need to learn enough to see what the big questions are," Alan says. "Where do you have an interest and think you can make an impact? Has your big idea already been published?"

But it's an exciting time now—new equipment and techniques are developing rapidly. This means there are opportunities for people to identify a niche in neuroscience. Alan draws from his decades of experience when he observes, "Tools from math, optics, solid-state physics, or electrical engineering can put you in a reasonably unique position to do work that few other people on the planet could do. However, remember that there are no guarantees. All you can do is work the probabilities and have fun doing it."

★ Now You Try!

Where Is Your Field Heading?

Sometimes we can get lost in the day-to-day of our chosen careers. It can be worthwhile to stop, take a step back, and imagine how your career, and the careers of those around you, will play out in the long term. Physical constraints such as costs, or even new inventions, can suddenly relegate entire industries to the past, even while new industries are being born. Don't make the mistake of thinking that just because lots of smart people are headed in a certain career direction, that it's what you should be doing, too. You may be in a good job now, but will it stay that way? Take out your notebook or sheet of paper and write the title "Foretelling Career Challenges." Then draw a line down the middle of the page to make two columns. In one column, outline possibilities for change in your area of expertise. In the other column, outline how you might successfully handle those changes.

Bonus exploration: If you are planning a major career shift, it can be wise to take tentative steps to test the waters—as Terry did when he started taking biology classes and enrolled in the Woods Hole neuroscience boot camp. If you *are* thinking of a career change, how can you test the waters and see if your projected career path might be the right one for you?

Chapter 9

Derailed Dreams Lead to New Dreams

It's difficult to be eighteen and find your dreams crushed.[1]

Such was the case with Princess Allotey.

Princess grew up in Ghana, in Klagon, near the capital city of Accra. Klagon is known for its high illiteracy rate and school dropouts. Princess's parents only completed a basic education—the equivalent of junior high school in the United States—but they had always encouraged their four children to go on to college.

English is the official language of Ghana, but most people from Ghana have both a Ghanaian and an English name and speak English as well as at least one of the seventy local African languages.

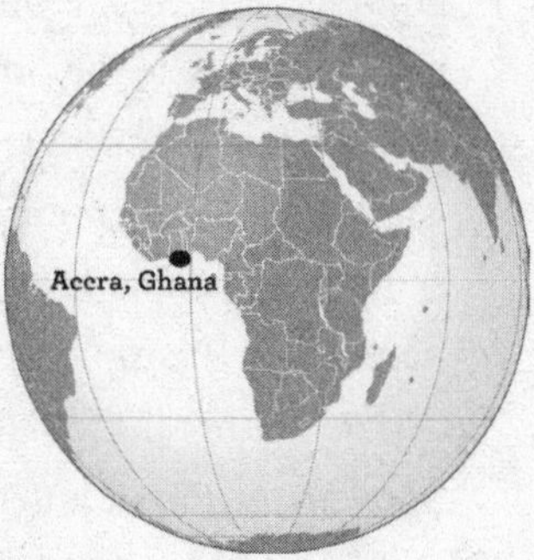

Princess was named "Princess" because she was the firstborn girl in her family—her oldest brother was named Prince. But since Princess's father was of Ga heritage, her full name, in accordance with Ga tradition, is Princess Naa Aku Shika Allotey. She speaks three languages fluently: English, Ga, and Asante Twi, the language of her mother, a Fante from Eshiem, a rural community in central Ghana.

In elementary school, Princess squeezed in with some eighty other children into a classroom meant for thirty. She and two friends typically shared one tiny desk. Despite the cramped conditions, she was hungry to learn—especially math—and she practiced a lot, asking her teachers many questions. She eventually earned A-pluses in math in both elementary and middle school. She also earned A's in the school certificate exams in all nine of the other fundamental subjects. This allowed her to be accepted into the prestigious Achimota High School—one of the best coeducational high schools in Ghana. Princess dreamed of one day becoming a math teacher—but not just any math teacher. She wanted to be a math teacher informed by insights on education from other parts of the world.

At age eighteen, Princess Allotey of Ghana started Kids and Math—an organization that provides school children with the basic mathematical resources they need to help them excel.

To help broaden her knowledge, she enrolled in a summer program to gain basic science, engineering, and technology skills and to become a more creative problem solver. She and her friend Shaniqua were the only girls among the twenty-one participants. Being different was hard on the girls—they felt the boys would be skeptical about whatever they might contribute. Princess felt like an imposter, despite the fact that the boys she worked with were supportive.

Young people typically have lots of dreams, and Princess was no different. For example, she longed to show the vision and courage of Nobel Prize winner Leymah Gbowee, who, in 2003, led women in a mass movement to put an end to the second Liberian Civil War.

The problem was that Princess had a terrible time speaking in public. It wasn't that she was shy—her conversation flowed easily when she was with her friends. However, if she even thought about getting in front of an audience, her nerves would kick in. Even with a written speech in front of her, she would find herself garbling words or she would just plain freeze.

The Achimota school Princess attended is supported by the older Achimotan alumni, including many of Ghana's ex-presidents and parliamentarians. It is also government-owned, which means that the fees are quite modest. Princess's father George, a hardworking man despite his struggles with asthma, owned a medium-size cement block company. He could easily afford to pay Princess's Achimota tuition and board fees. In spite of her challenges with public speaking, Princess excelled at her studies.

Until disaster struck.

Feeling Like a Fraud

"Imposter syndrome" is the feeling that you are not truly deserving of your accomplishments—or, at the very least, far less able than those around you. Though it's called a "syndrome," feeling like an imposter is not a mental disorder—it's simply an emotionally harmful way of framing your achievement. If you are successful, you think it must have been an accident or lucky timing. Or maybe people were somehow fooled. In other words, the way you see it, your success wasn't really your doing. If you fail, on the other hand, you see it as your fault.

Women, in particular, seem to experience these feelings often, though men can as well. (It's possible that men just aren't as forthcoming about their feelings.) As Drs. Pauline Clance and Suzanne Imes noted in their original research paper on the topic back in 1978: "Despite outstanding academic and professional accomplishments, women who experience the imposter phenomenon persist in believing that they are really not bright and have fooled anyone who thinks otherwise." And, sadly, this belief—this feeling that they are frauds—persists even in the face of solid evidence of their intelligence, achievements, and capabilities.[2]

Imposter syndrome, oddly enough, occurs most commonly in high achievers. Part of the challenge in overcoming the syndrome is that the imposter's humility can be refreshing for regular people who catch a whiff of it (*she's humble!*). Women, perhaps because of their heightened sensitivity to the feelings of others, might tend toward being bashful to avoid the stigma of being considered a braggart.[3] Testosterone may play a role here as well—the hormone is associated with aggressiveness, dominance, and risk-taking behaviors.[4]

Princess found herself experiencing full-blown imposter syndrome in the technical summer camp she attended. She was placed in charge of an all-male team designing a long-term vegetable storage container for farmers. In managing her team, she not only had to speak in front of the group—always a problem for her—but she also had to tell the group what to do. *Who was she to be put in such a position of authority?*

This "I am not worthy" attitude led Princess to take care in how she issued directions to her team. "Do you think this is correct?" she would wonder aloud. To her amazement, she began to realize that the team saw her as a leader—one who made good decisions. This encouraged her to take her blinders off and really look at what was going on around her. This informed, more objective assessment of reality, as it turns out, is an important step in overcoming feelings of imposterhood. Ultimately, Princess's reframing toned down those self-critical, second-guessing-herself circuits that were running in her head. Clearly she *was* capable. And she was further reassured of that by guidance from mentors. She began to realize something else—that she didn't always need to be stereotypically dominant, just ordering people around, to be a good leader. This, in turn, allowed her to recognize that she could move past the imposter syndrome while also benefiting from it.

Self-doubts are by no means always bad. Military officers and embassy officials, for example, can be full of a subconscious cultural rectitude that their perspectives are correct—an attitude that can get them into trouble once they arrive at their overseas assignments. In the realm of science, Nobel Prize–winning neuroscientist Santiago Ramón y Cajal said that one of the biggest challenges of the geniuses he worked with was that they would jump to conclusions and then be unable to change their minds when they were wrong.[5] History is replete with business executives, generals, and politicians who only listen to others when they reinforce their own thoughts—these leaders then steer with blithe conviction toward disaster. Doubt, of course, can be overdone—but it can also be undervalued.[6]

And the reality is that although talents and skills matter, luck can also play an important role in our lives. A toss of the dice between two equally talented applicants can leave one with a job and the other feeling like a reject. A concussion from an out-of-the-blue automobile accident can mean college prep exams go badly—meaning your chances of getting into that top university are reduced. Perhaps the most wonderful piece of luck of all is to be born into a loving and supportive family—a sort of luck that some can only wish for.

Thus it's natural that most of us—except perhaps the most brash and narcissistic—can occasionally fall prey to feeling like an imposter. Accepting that these feelings are normal and reframing them to our advantage forms a healthy way to move forward.

Princess's Problem

Princess was highly focused on her studies at Achimota High School—in that elite environment, she averaged a 3.7 GPA. Meanwhile, Princess's father, George Allotey, was facing a good problem: his business was doing so well that he needed to expand. To do that, he bought additional land, paying more than 300,000 Ghanaian cedis (Ghc) in cash—about U.S. $75,000. In Ghana, that was an enormous sum. The individual he bought the land from was powerful, highly placed, and (George thought) a longtime friend. No receipt was given—George never told his family why, but perhaps he felt embarrassed requesting written documentation from such a powerful mentorlike figure. In any event, George knew that once he started building on the land, it would be clear that the transaction had been approved.

But before George could get a start with building, disaster struck. Another businessman disputed George's claim, saying *he'd* purchased the land.

In any legal case, it is often difficult to tease the truth from claims and counterclaims, and of course, this story is told from Princess's perspective. But in circumstances such as these, there can be a compelling impetus for a landowner to accept money from two parties. A single decisive word from the landowner could settle an issue like this in favor of the proper claimant—but then, the landowner would receive only one payment.

There was no decisive word. Instead, the landowner simply suggested that the two parties settle the matter in court.

Because George had poured such an enormous sum of money into the location, he couldn't afford to just walk away. Beyond that, he was a very determined, hardworking man—which is why he had been so successful in the first place.

During Princess's final two years in high school, the litigation ran its course. George was forced to travel back and forth to the capital to file court documents, all the while paying enormous sums to his lawyer. He was desperate—he'd paid for the land, after all. But when he meanwhile tried to start building on the land, he and his crew were dragged off and beaten. One of his workers ended up in the hospital and George's expensive tools were destroyed—some even looked like they'd been run over by bulldozers.

This strife began to take a toll on the family and on Princess's studies. She continued to work hard, but her grades began to fall. Sometimes, heartbreakingly, she even got D's—unheard of for her.

Still, Princess kept working, trying to do her best. As the final year of high school rolled around, she began studying for the critical West

African Secondary School Examination, which would determine whether she could get into a university—whether at home in Ghana or abroad. The test would begin at the end of February 2014. She was determined to give it her best.

George had borrowed Ghc60,000, about $15,000, from a well-to-do friend to take his fight for the land to the highest courts. He had put his all into the case, and, as the proceedings rolled on to completion, he was down to Ghc250—about U.S. $60.

On January 2, 2014, two years after the litigation started, George learned the court's decision.

He had lost.

The next day, George's asthma worsened dramatically. He sent Princess off to the pharmacy to get drugs. While she was gone, George passed out in the kitchen, where he was found by Princess's mom. She struggled to get him into a taxi to the hospital. The first two drivers declined, refusing to take what they said was a corpse. The third driver accepted them into his car and sped to the hospital, but to no avail.

George had died.

And then, just when the family thought things couldn't get any worse, they did just that. As it turned out, George had been so convinced that he had a winning case that he had used the cinder block factory and the family's house as collateral for the Ghc60,000 loan.

The family hadn't just lost the court case. They'd lost pretty much everything they had.

Two months later, Princess took the all-important West African Secondary School Examination. Remarkably, her results were outstanding.

But Princess had no money, and lacked the connections she would need to get a scholarship, so attending full-time university in Ghana was out of the question. Her applications to overseas universities came back with acceptances—but no support monies. As it was, the family was scraping by on the slim earnings of her oldest brother.

More rapidly than Princess could have imagined, she went from having her dreams within her grasp to discovering that her options had slipped away.

Reframing—and Developing New Talents

In previous chapters, we've learned about the vital importance of reframing, which, for example, Adam Khoo uses to see problems as opportunities.

Princess, too, discovered the value of reframing. Her final years in high school had been hard on her. With her father's death and the financial ruin of her family, she'd fallen into a depression and her grades had suffered. However, she'd picked herself up as best she could, and she performed exceptionally well on the final high school examination. For Princess, it was her religion that helped her reframe. Her family is Catholic, and she found that her faith and the values that go with it—such as the call to aid the less fortunate—were supportive in getting her through the difficult times.

Instead of ruminating on her own problems, Princess reframed her thinking by looking outside of herself—thinking about how she could help others with *their* problems. She began volunteering as a teaching assistant in less-endowed elementary and middle schools. Here, among impressionable children, she shared her excitement and enthusiasm for math. She wanted to help all the kids, but she especially wanted to set an example for girls. In Ghana, it's more acceptable for boys to be into math than girls—in fact, it's still more acceptable for boys to attend school.

Most of the kids she taught couldn't afford math books, which made it difficult for them to study, practice, and reinforce what they were being taught in school. So Princess came up with Project Arithmas, to put together a library of math books at the school to help the kids prepare for their Basic Education Certificate Examinations. She enlisted eight friends and ginned up a budget of Ghc700 (about $175), which they scraped together from their own pockets and from a few kindly supporters. Their plan was to purchase a few key books and ask authors of various math books to donate copies. (This is how I came to meet Princess—she wrote to me to request a copy of *A Mind for Numbers*. What especially impressed me, however, was her follow-through. She didn't just write to get the book—she wrote to express her appreciation once she'd received it. During my own previous visits to Africa, I'd seen firsthand the challenges experienced by African school children—Princess was tackling those issues in one of the most direct ways imaginable.)

Princess went on to found and become the executive director of an organization called Kids and Math, to encourage appreciation of mathematics. Her work required travel to many different schools, where she talked to the kids to excite them about math. To raise funds for Kids and Math, she gave speeches to various agencies, companies, and groups. She also coordinated fund-raising through the sale of trash bags. She bought the trash bags in bulk, and sold them individually at a slightly higher, but very affordable price of Ghc0.80 (about U.S. $0.20). Such bags, which were customized for medium and small kitchen-type dustbins, are rarely sold in Ghana—customers bought them not only because the bags were attractive and handy, but also to support Kids and Math.

Princess has become an entrepreneur—a social entrepreneur. That is, she now uses the techniques of business to solve social problems. And, to her amazement, she has become something else—step by step, presentation by presentation, she has become a polished public speaker. When she and her team were invited to give a speech on Kids and Math at a meeting of the Toastmasters Club at Ghana's Ministry of Foreign Affairs and Regional Integration, Princess saw the opportunity to get an evaluation of her presentation skills. After the club meeting, a Toastmaster congratulated her, saying, "Your speech was great—the presentation was just like a TED talk!"

Princess began to be get noticed. She was asked to do an interview about Kids and Math on the popular television show *GH Today*, co-hosted by popular star Kafui Dey. When she went on the show, she kept wanting to see if someone else was behind her—after all, she herself couldn't really be speaking with Kafui Dey!

She had to laugh when she realized she felt like an imposter.

The interview went beautifully.

Princess's ability to reframe and see challenges as opportunities hasn't (yet) allowed her to achieve her dream of getting a university degree so she can get formal training in being a math educator. But her reframing has done something else. It has given her a powerful sense of purpose. It's allowed her to overcome her feelings of being an imposter. And coincidentally along the way, she's managed to overcome one of her greatest challenges—learning to be a public speaker.

In this chapter, we met a young woman with a love of math who overcame her feelings of imposterhood, whose dream included a highly nonanalytical skill—public speaking. In the next chapter, we'll meet a techie type who has escaped high tech altogether.

 **Now You Try!**

Embrace Your Inner Imposter

Do you sometimes feel like an imposter? Do you feel as if others in your situation are somehow better than you, and by comparison, you're a bit of a faker? If so, you're not alone. In fact, you'd be surprised at how many secretly feel the same way you do, even though they may put on confident airs. (Sometimes overly confident—the person who told you, for example, that he aced the midterm might have in reality only gotten a C.)

Feeling like an imposter can lead to discomfort and doubt, but is not all bad. It can help you to look with the dispassionate eye of an observer at what's going on around you. It can also help you avoid the arrogant overconfidence that can lead to bad decision making and poor leadership skills.

Take out a piece of paper and, below the word "Imposter?" with one sentence at the top, describe a situation where you are feeling like an imposter. Below your sentence, draw a vertical line to make two columns. On the left, write about the positive aspects of feeling like an imposter. On the right, write about the negative aspects.

Then write two or three sentences (or more) to synthesize your feelings about imposterhood.

Chapter 10

Turning a Midlife Crisis into a Midlife Opportunity

ARNIM RODECK HAD known he was going to become an electrical engineer ever since he was a kid sitting in his bedroom tinkering with electronics. He never anticipated that he would grow dissatisfied with the job he loved—a job that he seemed born to do.[1] And he certainly never foretold the turn his career would take.

Finding Paths Past the Insurmountable

Arnim was born and grew up in Bogotá, Colombia. His sweet-natured, supportive mother—a nurse—was born in Africa to German and Belgian parents, while his strict, results-oriented father was an Austrian from Vienna who owned an elevator business. The couple had fallen in love in Colombia and ended up staying there, so Arnim grew up comfortably bilingual, speaking Spanish and German. He jokes that he likes to use the German area of his mind for more logical thinking and the Spanish area for the more emotional and passionate parts of his life.

The school Arnim attended as a youngster was partially sponsored by the German government, so some subjects were taught in German,

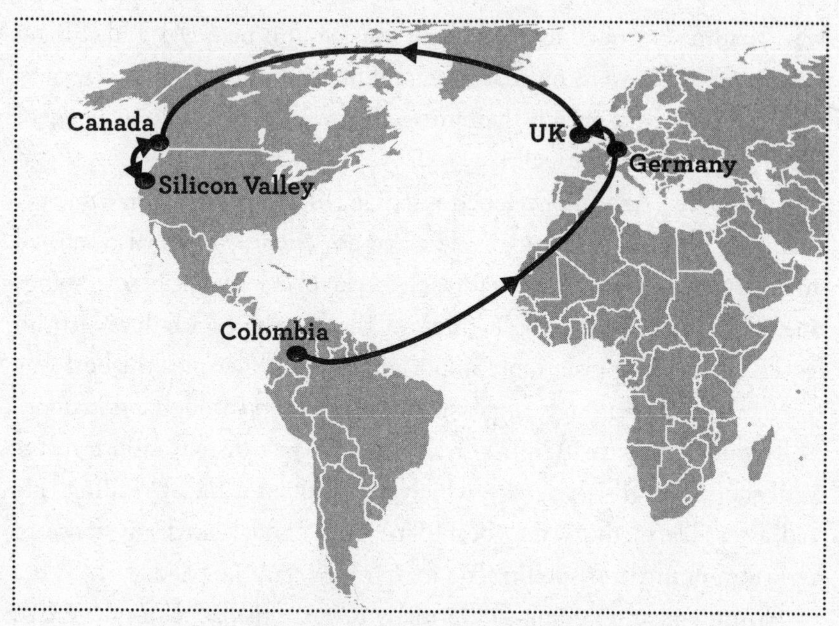

Arnim Rodeck's learning journey took him far from his native Colombia before he found his career calling.

some in Spanish, and a few in English. Since Arnim's first English teachers were Germans, his English has a German accent.

But it wasn't all roses for Arnim. He is dyslexic, and memorization has always been difficult for him. This meant that schoolwork was a struggle. And he had other challenges—with music, for example. He's terrible at singing—so terrible that, in kindergarten, when it was time for all the children to sing, his teacher had him go play with Legos.[2]

Arnim also had no rhythm, so he couldn't dance. He was unable to read musical notes or detect which piece of music he was hearing, and he couldn't even figure out which instruments were doing what.

But audio signal processing—the digital and analog aspects of the electronics behind some forms of music—was a different experience from hearing music for Arnim. He *loved* it. Fortunately, his high school music teacher recognized Arnim's hidden strengths and allowed him to pass his exams by building a turntable and an electric guitar. In this way, Arnim continued to learn about music in his own way and still get great marks. He went on to design and build music synthesizers, mixers, recorders, and even a theremin—that weird electronic instrument that's played without touch.

Ultimately, Arnim developed a strong, lifelong interest in a subject that most people would have said he had no "talent" for. He also learned an oft-overlooked lesson: Great teachers can bring out the best in someone, even when others might think of that person as a failure. Arnim learned something even more important—that sometimes the best way to succeed in a seemingly impossible task is to slip through a side door.

Thanks largely to Arnim's dyslexia, he was a dismal failure in his high school English classes—which necessitated a lot of reading. He had a terrible memory for vocabulary words and lacked any sense of logic for grammar or spelling, no matter how hard he tried.

Arnim ended up going to Germany to earn his bachelor's degree in electrical engineering. To his surprise, once he arrived in Heilbronn, he found himself facing his old nemesis—the English language. Some of the required technical courses were taught only in English. Tests were in English, too. So he struggled valiantly—passing exams only with the help of tutors and the occasional kind, blind eye of his professors. He was warned to avoid pursuing any jobs requiring English.

However, much like Singaporean entrepreneur Adam Khoo, Arnim had learned to turn apparent disadvantages into advantages. After his move from equatorial Colombia to temperate Germany, he realized that he loved studying in different countries, where he had the chance to meet diverse people and soak in new cultures. For graduate school, he ended up in the UK, language "disability" and all.

To Arnim's surprise, he discovered that despite his problems with *reading* English, he was quite good at picking up the spoken language. He had no language-related problems in his master's studies at all.

> Though my English was very poor when I got to England, I never have been shy, so I simply asked questions and spoke, regardless of how bad

I was. The fact that I was a foreigner was often a great excuse to ask for directions, things to do, special places, and so on—questions that a natural English speaker from the area often wouldn't ask.

For example, while looking for an appropriate university for my master's, I was in an almost empty train going from Manchester to Liverpool. I sat beside a young woman and started chatting with her. I wondered aloud about the various master's programs and finally asked for a recommendation of where to stay in Liverpool. She ended up inviting me to her parents' home and became a wonderful friend and supporter.

Having an accent makes people curious and often gives you a chance to tell them a bit about you. It breaks barriers. But the biggest advantage of speaking more than one language is simply understanding that there is more than one culture. There are several ways of seeing and acting in the world. Learning to speak another language gives you a more open mind.

I think the language-study turnaround for me was moving from the structure of a formal class to simply speaking and interacting with people. As a matter of fact, even today I keep diligently learning new words that I pick out of books and the news—I practice them every morning through a flashcard system called Anki. Ironically, learning new vocabulary was what I hated most and was terrible at in school. I am still slow, but now I simply do it and even enjoy it.

In a funny way, my bad English even allowed me to communicate better. People made an extra effort to understand what I was saying. They were interested and helped me learn by being unwilling to accept language as a barrier.

Distraction: It's Not Necessarily a Bad Thing

It turns out that listening with a bit of a distractor—like a foreign accent—gets your brain to do mental tricks to boost your thinking. So Arnim's insight that people might have paid closer attention to him because of his accent was spot-on. When people have slightly more difficulty in processing what they are hearing or seeing, they can be forced to think more abstractly. This, in turn, may allow them to think more creatively about what they are hearing.[3]

A bit of background noise, much like having an accent, seems to create slightly more difficulty with processing—it distracts just enough so that you move, at least momentarily, into a different mode of perception, allowing you to think more broadly and creatively. This may be why some of us go to a coffee shop, with its background murmuring—perhaps subconsciously seeking an obliging ambience for our studies.

Focus Is Good, but Not All Learning Involves Focusing

When we're studying we often drink caffeine, which enhances focus by attenuating "daydreamy" alpha waves. This effect is strongest for about an hour after drinking a cup of coffee or tea, although the energizing can persist for some eight hours, which is why an evening cup of coffee is sometimes better left untasted.[4]

When you're doing something that's cognitively difficult, coffee isn't the only booster. You often subconsciously use other tricks to increase your focus. For example, if you're trying to remember something, you tend to avert your gaze—which avoids overloading your working memory with extraneous information from your environment.[5] Even just closing your eyes can help you ignore distractors so you can more easily bring something to mind.[6] Accordingly, memory experts who compete in memory competitions will do everything they can to reduce noise and extraneous visual stimuli, often wearing special blinders and earmuffs so they can stay focused.

It's often easier to just memorize something than to truly understand it. This sometimes trips up those medical school students lucky enough to have exceptional memories. (Yes, although memory tricks work well, memorizing does come much more easily for some people. Researchers still aren't quite sure why, though there's evidence that having the right genes helps.[7])

In medical school, when there's a big anatomy test, ordinary med school students spend weeks preparing. They practice over and over to memorize thousands of terms and related functions. Ace memorizers, on the other hand, can procrastinate until just a few days before the test, spend a few hours glancing over the material, and still do well.

However, when these same memorizing aces are faced with a different type of medical school exam—for example, a test related to how the heart functions—they find that a few hours of last-minute cramming just doesn't cut it. Medical school advisers themselves can be startled to find these seemingly star students flunking certain sections of the curriculum. It seems that quickly memorizing anatomical terms related to the heart doesn't allow you to understand and answer questions about the heart's complex function.

This is a reminder that simple focused concentration often *isn't enough* when we're trying to understand a complicated issue.

More Complex Learning Demands Diffuse Connections

It takes time to understand complicated systems, whether we are studying the human heart, laying out a new lawn irrigation system, or analyzing the multifaceted causes of World War II. To untangle such complex subjects, we often need to alternate a tight focus on the issue at hand with steps back to look at the bigger picture. Our need for occasional distraction during any given learning session may arise from those competing tight-focus versus big-picture needs.

As you saw in Chapter 7, people have two very different ways of perceiving the world—that is, two different neural approaches to thinking. The focused mode makes use of our focused attention, while the diffuse mode involves neural resting states.[8] Focused thinking, if you'll remember, is the kind of thinking you do when you are intently concentrating on a math problem. You might fall into diffuse thinking, on the other hand, when you are standing in the shower not thinking of anything in particular.

Now, let's take these ideas a bit further.

The focused mode is primarily centered in the prefrontal cortex—the front part of your brain. The diffuse mode, on the other hand, involves a network that connects more widespread areas of the brain.[9] The widespread nature of diffuse thinking is why it is often related to the unexpected connections that lie at the heart of creativity.[10] Activities involving the diffuse mode, like walking, riding a bus, relaxing, or falling asleep, are more likely to lead you to creative ideas that arise from seeming nowhere.[11]

A Little Background Noise

If we're in a very quiet environment, that quietness can hype up focused-attention circuits while simultaneously deactivating the diffuse mode. This is why quiet environments are ideal when we're doing something that demands concentrated attention, like our tax returns or a difficult problem on a test.

But sometimes we're looking at bigger-picture sorts of issues—like cardiac function or computer network connectivity or meteorological patterns. In that case, a little sporadic noise, like a snippet of conversation or the clatter of dishes in the background, can help. This is because that bit of noise temporarily allows the longer-range diffuse network to pop up. (Technically, the noise "disrupts the deactivation of the default mode network."[12]) In other words, the gentle hubbub of, say, a coffee shop can still allow you to focus, but the background noise also allows you to step back more easily on occasion to look at the bigger picture of what you are trying to understand.

But there can come a point when there's too much noise, which can keep you from concentrating at all. Older people can be a little more sensitive to noise, because they're not quite as good at suppressing the default mode.[13] That's probably why older patrons at restaurants tend to take more issue with neighboring conversations that are all trying to outcompete the evening's musical entertainment.

Key Mindshift
Background Noise

A little intermittent background noise can allow us to more easily alternate between focused and diffuse modes. This is especially useful with learning that encompasses new concepts, approaches, or perspectives.

What About Music?

So, you might ask—what about music? Does it help or hurt when you are trying to study? The answer is—it depends. If the music is fast and loud, it definitely disrupts reading comprehension, in part because you use some of the same areas of the brain to process music as you do to process language.[14] Music with lyrics is more distracting than music without lyrics.[15] On the other hand, researchers have found that if you're listening to a favorite style of music, it could enhance your studies, while if it's something you're not so fond of, it could detract.[16]

In the final analysis, all this just means that when it comes to music, you should use common sense and discover what works best for you.

Another Slip Through the Side Door

Arnim's Colombian upbringing taught him a mind-set rather different from that of people in many wealthier nations. The country of Colombia is not just developing—it is *rapidly* developing, and its diverse peoples have a confident, enterprising spirit. If homework is due and there is a power outage, teachers still expect assignments to be turned in on time, no excuses. If there's horrific traffic that means a simple trip across downtown Bogotá could take three hours, it doesn't matter—the work still needs to be done on time. The intrepid cultural expectation that he could find a way around obstacles was part of Arnim's very soul.

In Germany, Arnim often heard, *So etwas haben wir noch nie gemacht*—"We have never done that." People who said this really meant that *Arnim* couldn't do it. But as soon as Arnim would hear that, the Colombian part of his mind would start wondering, "Well then, how can *I* do it?" He used this thinking to avoid having to retake all the courses he had already taken at a previously unrecognized university in Colombia. When he asked the dean about how he could obtain formal credit for his previous coursework, the dean first replied, "Can't be done"—but then added, "Well, unless you get an okay from every single one of your professors."

Arnim asked around and discovered who the "easy" professors were. He got their signatures, gaining a momentum of yeses, so that, eventually, even the toughest professor couldn't turn him down. In the end, the dean congratulated Arnim and gave him credit for his previous work.

Upon graduation with his master's, Arnim found himself in a quandary. Although he'd always wanted to move to Canada, he hadn't had much luck in pitching himself for jobs there. As graduation crept closer, he sent out hundreds of applications to companies in Germany, also with little success. It was depressing. He went to engineering-telecommunications job fairs and saw massive lines at all of them. He couldn't resist asking the human resources people he spoke with whether they knew of any less popular, less crowded job fairs.

Well, off he went to the "wrong"—but far less crowded—job fair, one that specialized in economics. There, he was able to speak to representatives from many of the same companies and countries as were at the engineering-telecommunications fair. Most gave him a hard time for turning up at an event meant for a different industry. The rep from Hewlett-Packard, however, loved his moxie, saying, "We're looking for people who think differently!"

Arnim was first hired as a support engineer for HP in Darmstadt, Germany. As part of his training, he was sent to an HP lab in Bristol, in the United Kingdom, where the new products were being developed. Here at last, Arnim felt he began to get his *real* education—often, through mentors.

Arnim's first mentor didn't talk a lot—he was instead an outstanding listener who led by example to get the best from people. When he did speak, he was onto something. Arnim learned from his next mentor not to worry about money, positions, or even reputation, but to focus on doing the best he could, without cutting corners.

Much like high school dropout Zach Caceres in Chapter 5, Arnim has found mentors to be extraordinarily important in his career and personal development. He's had "professional mentors" who were paid by HP as part of a future leaders program. But it was his self-appointed mentors who have made the bigger difference.

When Arnim spots a potential mentor, he works to get their interest. He realizes, for example, that a lone e-mail isn't enough. He's come to realize that different approaches appeal to different people—there's no single "mentor-grabbing" trick. He also recognizes that it's a big turnoff to ask someone point-blank to be your mentor, especially when they hardly know you. Again, like Zach Caceres, Arnim looks to see how he can make the relationship a win-win, so the mentor will also gain through his or her "investment" in their relationship. Arnim also makes a point of seeking two different types of mentors—one who keeps him feeling confident and pumped up, and another who doesn't shy away from criticism and who broaches no excuses.

Key Mindshift
Mentors

Mentors can be invaluable in your career and personal development. People don't need to even know that you consider them to be mentors for them to be of value in your life. Look for ways to make yourself somehow useful to the mentor, just as they are for you, to make such relationships flourish.

Arnim's trip to the world-class laboratories in Bristol also provided him with an opportunity to show his mettle. The project Arnim was assigned to had special challenges. Oddly enough, some of those challenges came, not directly from the technical problems, but instead, from the lab's culture. In old-fashioned British stiff-upper-lip style, one simply didn't ask for help. But perhaps due to Arnim's being a foreigner in addition to being an inexperienced newbie, he *had* to ask around to find out what was going on. Executives began to notice Arnim's willingness to reach out, form connections, ask questions—and get problems solved.

HP had just bought a small start-up in Canada, and its management was looking for a support person in Europe who was open-minded, flexible, and willing to build HP's presence. Arnim fit the bill perfectly. But once Arnim began visiting Canada, he became engrossed in resolving problems that came up at that location—his trips became more and more extended. After a year, it started to seem silly to keep his apartment in Germany. His dream of moving to Canada had come true.

It was Arnim's love for new cultures that next led him to one of his biggest breakthroughs—selling a major customer on a new product that had never previously gotten any traction with the company. Arnim convinced his managers to let him go into the field in Silicon Valley and "live with" the customers. His move was very unusual, because the company had a well-established sales and support force. However, within six months, Arnim began to understand how customers were using both HP's and other companies' products. His reports back to the factory formed a new dialogue—a way of listening to what was truly needed—that enabled HP to become more successful. And that's how he came to find himself living in the hotbed of high tech, in Palo Alto.

Time to Change Careers

It might have been Arnim's father, Heinz, who planted the seed for career change. He would always say that when you become good at what you do, it's time to change. "Don't wait for what comes after that." He meant, *don't wait until you're tired of it.*

It wasn't burnout, exactly, that Arnim experienced. For one thing, he was treated well at Hewlett-Packard (part of which was eventually spun off to form Agilent Technologies). He was valued, he had great colleagues, and he felt intellectually challenged. It was just that he was tiring of the politics and bureaucracies that are inevitable in big organizations. He was also tired of the daily maddening commute through traffic jams, the vistas of buckled concrete, and the world-class, yet sometimes surprisingly narrow-minded, people around him whose interests centered solely on technology or business.

And so, after more than a decade doing his near-dream job, Arnim began thinking about a career transformation. He had no idea what his next job would be; he just wanted to become good at more than one thing, and he wanted the new job to be different from what he had been doing. Of course, change entails risk, but the risk of *not* changing could be bigger.

Underlying his need for change was the fact that Arnim wanted to be his own boss—and to be a creator. He also wanted to find something that he could get better at, even as he grew older. One of his strongest assets was his analytical way of thinking, arising from his engineering training and experience—he knew whatever career he chose should take advantage of that.

Gradually, he began thinking about options. He kept a piece of paper with him and jotted down any idea that came to mind—especially the crazier ones. At the end of each week, he took the pieces of paper and sorted through them. After six months, one idea was shining through.

Woodworking.

Arnim hadn't done any real woodworking before. But he was drawn to the beauty of the woods in Canada and inspired by how native carvers breathed unique life into each piece. He loved the feel of wood—how he somehow communicated with a piece, and it with him, signaling how the woodwork should unfold. It was the opposite of what Arnim had experienced in tech, driven as it was by emotionless consistency, accuracy, and efficiency. "With wood," Arnim says, "it comes down to feelings, perceptions, patience. In essence, art. I wanted to explore that side in me. I wanted to make a new career and, with that, a new way of seeing the world."

Once he'd figured out what he wanted to do, Arnim began visualizing himself ten years into the future, working in his wood workshop with his customers. Next, still imagining himself in the future, he would ask himself, "How did I get here?"

Two things became clear. First, he *loved* what he saw. He wanted to go there.

Second, he would have to quit his job—undo his "golden handcuffs" and jump into unfamiliar waters. Of course, this would not guarantee success.

The advantage of Arnim's "envision the future" method was that he didn't have to plan all the steps. Instead he simply geared his mind toward building a new career that incidentally made use of his previous engineering training.

When Arnim finally quit his high-flying job, many of his colleagues thought he was making a big mistake. But at the same time, they envied him. Since then, some of his old coworkers have visited and even enjoyed helping him in his workshop. (A fair number of those old colleagues have since lost their jobs in mergers and high-tech shuffles.)

The shift was hard—much, much harder than Arnim had thought it would be. Since he had no real previous woodworking expertise, he needed to learn the best techniques and to experiment with woods, glues, and finishes. He also needed to learn where to get the best materials, and how to keep himself up-to-date with key professional developments.

Arnim had never even run a business before. He had to figure out what he should sell and to whom. He had to determine costs, location, and logistics, and understand all the intricacies of money flow. He began to realize that he'd been spoiled by his past experience at a big company, where there were specialized departments to take care of all of his and the company's needs.

A good bit of Arnim's challenge came down to prioritizing where to put his energy. There were many more problems and challenges than he had time for. He had to figure out how to do it all himself—advertising, selling, procuring, shipping, testing, building, answering requests, designing, fixing problems, experimenting, and meeting new clients. And then there were building codes and other regulations.

In a way, not knowing at the beginning how hard it was going to be helped Arnim to keep sailing through sometimes turbulent waters.

A key aspect of what Arnim did was to begin to visualize himself ten years in the future. He could see himself in his wood workshop with his customers, imagining what his life looked like. He loved *what he saw.*

As Arnim has come to realize, in his old "envisioning the future" mode, he was not able to plan the details of a profession he didn't yet know, or the details of running his own business. But those original dreams began programming his "vectors"—his subconscious thoughts—to keep pointing him in the new direction. This subconscious vectoring hasn't stopped, of course. Even now, Arnim keeps envisioning how he needs to keep changing, learning, and growing—augmenting his current skills. He says, "If I have a free moment, you will see me in the workshop dreaming up the next thing. Wherever I go, whatever I do, I am dreaming of the next things I want to do."

He has worked hard to build an environment that forces him to keep changing—to not be content to fall back into old ways. Now, more than ten years later, Arnim is more passionate than ever about his new career—even though much of it did not turn out as he had initially envisioned.

Arnim had loved the people he'd worked with in industry—highly intelligent people who kept him on his toes. To keep the spirit of these people with him, Arnim began to take notes on people he admired, respected, or even disliked—people who were good at what they did. He observed what types of questions they tended to ask and what made them so good.

Nowadays, phrases from Arnim's old "guiding light" mentors at HP still drift through his head, keeping him on track as he does various woodworking projects. Some of these phrases are:

- Yes, it's got a lot of great features. But how well does it do on the basics?
- *Be* the customer, *use it* as the customer, try to *achieve* what the customer needs to achieve with it.
- Ensure there is a win-win-win for everyone—the supplier, the customer, and us.
- Know and focus on your talents and successes. But also know your deficits and have ways to get help for these.
- Think about the future. Each step you take, as small as it might be, adds up and becomes powerful down the road. Think of compounding interest.
- There is no such thing as a problem with a customer—it's just an opportunity to establish a deeper relationship.
- You don't take a sales course—or anything else—and think you know this stuff. Keep doing it for a decade and *only then* you will start getting it.
- Find out how to get the best out of people and help them thrive. It will make you thrive, too. Not the other way around.

Even today, Arnim sometimes imagines meetings with his old HP colleagues to keep their inspirational approaches alive within him. "I recall their questions and attitudes and review my latest idea or challenge with 'them.' Of course I cannot replace working with those people, but I have taken some of the best of them with me into my new environment."

★ Now You Try!

Creating Words of Wisdom

Arnim made a list of the most useful pieces of advice his coworkers had given him. In a similar vein, who have you admired, respected, or even disliked? Who are the people in your life who are exceptional at what they do? What kinds of questions do they ask and points do they make?

Under the title "Words of Wisdom," make a list of sayings from your own favorite (and "un-favorite," but highly competent) coworkers. Your choice of words of wisdom should be shaped by your desires and goals—which will mean that you yourself have creatively contributed to your list. You can refer to this list as a guide in your plans for the future.

Retraining

One of the things Arnim did *not* want to do as he changed careers was to formally retrain. He instead wanted to nurture his free-thinking creativity, which was part of what had drawn him to woodworking. So instead of enrolling in a lengthy formal course, he took short courses and worked and studied independently—reading books and visiting woodworking-related fairs and shows where he asked lots of questions. He also tested ideas with potential customers to get their feedback, and worked on various renovation projects in his own home to get a better sense of his abilities.

Then a special opportunity arose. While visiting with friends and family back in Colombia, he heard about, and then went to, a religious retreat at a woodworking monastery in the town of El Rosal, near Bogotá. Arnim isn't religious himself, but he has respect for people who devote their lives to something greater than themselves.

Among the monks' leadership was a silver-haired German "Meister" woodworker—a much-beloved man who seemed to come straight from a medieval woodworking guild. He epitomized the magic that was manifest in the team. Under the Meister's direction, a team of twelve local carpenters made wooden items for churches, prisons, and people who commissioned their work. Arnim asked the Meister if there was any way he could just sweep the floor or otherwise clean up after his woodworking team—anything so that he could just watch for a period.

The monk, a gentle, soft-spoken man, gave Arnim a brief, noncommittal answer.

When Arnim was back in Canada, he wrote to the Meister—sending him a letter (the old-fashioned kind, with a stamp). No response.

Next Arnim turned to the telephone. The Meister, perhaps sensing he could trust Arnim, and possibly also impressed with Arnim's tenacity, simply replied from his end: "You are always welcome."

These were the magic words Arnim had been waiting for.

"How long should I stay?" Arnim asked.

"That is up to you" was the answer.

What Arnim had asked for had no precedent—like all the times before when he had asked for the impossible. No one had ever been given carte blanche to study briefly at the monastery—the expectation had been years of apprenticeship.

Arnim coordinated a fourteen-day stay. He lived in the monastery, ate with the monks, and spent his days doing carpentry. He was living his dream—it was one of the greatest experiences of his life. He made every moment count, asking questions and trying to learn from everyone, but also trying to be modest, appreciative, and helpful in every conceivable way. He studied in the monks' library and took notes, which he shared with the monks. While he was there, he began building his first projects, always seeking criticism and feedback.

Arnim's enthusiasm was contagious. The monks and carpenters were inspired by their protégé's obvious respect for them and the work of the monastery. They also appreciated his ability to learn quickly. Even today, Arnim regularly returns to the monastery to show the team how their training and ideas continue to inspire him. They sit, laugh, exchange ideas, and inspire one another. When Arnim is back in Canada, each time he reviews his old notes, he sees even more in them.

"Perhaps what was most important to me," Arnim recalls, "was how the Meister encouraged my learning. To observe and then try it out myself. To observe again and try again—to keep trying until it was *beyond* what I thought could be done. And to do it so often that it became innate." The Meister was trying to ensure that Arnim developed an attitude of continual improvement—to make sure Arnim didn't become complacent at a plateau. Arnim still hears the Meister's voice in his head every day as he works in his workshop.

In keeping with the Meister's guidance, Arnim makes certain that when he gets new commissions, he is continually pushing himself in new directions. With every project, he tries to ensure he has the chance to learn not just one new approach, but many. Arnim has crafted doors and furniture for the Vancouver Winter Olympics, as well as for many high-end houses, along with coffee tables, wall carvings, gift boxes, music stands, mantels, cabinets, street signs, and even, if it catches his fancy, commissions for items as simple as cutting boards.[17] And his customers have become his friends.

Energy Injections

Arnim's old environment in high-tech engineering was very energetic, with lots of insightful conversation, ideas, and coworkers motivating him. How, he wondered, could he maintain the energy that arises from working in a team when he went solo?

Considering this helped Arnim realize—counterintuitively—that time was *not* his most important resource. In fact, it was energy—both physical and mental. How could he develop and maintain it? Arnim started to do a lot of walking, hiking, and bicycling. He began to notice that interesting ideas and solutions to problems came when he walked in nature. Even his post-walk showers paid off: "The shower is my creativity office!"

Arnim hadn't realized how much he'd miss the performance evaluations he'd gotten at HP. He wasn't thrilled by negative feedback, but he had always used the evaluations to improve. Nowadays, Arnim instead conducts what he calls "postmortems." He does one after every project, asking himself—and his customers, friends, and colleagues—lots of questions so he can understand how to improve the next project.

Arnim, who takes great pride in meticulous craftsmanship, also makes an effort to experiment and to embrace his mistakes. He knows that creativity requires an openness to mistakes. As he puts it, "When things go wrong or not as anticipated, it's fun to take a positive approach and still find a way to make things turn out even better than expected."

Turning Negatives into Positives

Many people are turned off by a negative experience with a teacher—for example, if they had a bad math teacher they might use that as an explanation for their failure from then on. Arnim is different. He makes an effort to view even the most disheartening teachers as mentors. For example, as Arnim was reaching adolescence, he took math from a widely disliked teacher who practically exuded nastiness. Once, the teacher called Arnim to the front of the class and told him to draw a big circle. Arnim did that. "No!" his teacher exclaimed, "Much bigger!" Arnim complied. His teacher turned to the class and announced, "That is Arnim's mark on the math exam!"

Arnim was devastated, but decided that this was not to be his destiny. Arnim's father had been offering to work with him on math, and this public humiliation by his math teacher finally pushed him to accept his father's offer. Decades later, as a successful math student who went on to obtain a master's degree in electrical engineering, Arnim has a surprising perspective: He believes his math teacher did him a favor by pushing him to accept that he really did need help with his math.

 **Now You Try!**

The Power of the Positive

Even the most disheartening people can make positive contributions to your life. Under the title "Turning Negatives to Positives," jot down ideas about how you can turn negative encounters into positive learning experiences. To add extra oomph to this exercise, reach out to an upbeat friend and exchange some of your upbeat ideas. (Don't let yourself fall into waxing about the negative!)

Risks and Change

Arnim took substantial risks to build a new career. In the end, however, the initial discomfort he endured was preferable to spending hours each day commuting, all the while wondering whether he might suddenly find himself made redundant or obsolete.

"The most interesting lives," Arnim observes, "are lived by people who take risks and make mistakes—and who are willing to learn from those mistakes." He explains that to him, "the gift of having a mind brings an accompanying duty to experiment with it, to mold it, to play with it."

As an electrical engineer, Arnim can't resist comparing his brain to an operating system. An upgrade in the operating system usually offers better features, though there are almost always temporary glitches and problems. Arnim feels he needs risk to give him the forceful push he needs to open his mind to change. And indeed, in shifting into his new career, Arnim had to change his mind, his attitudes, and his values.

But Arnim found one way of changing that was more powerful than virtually anything he'd ever tried. That's what we're going to explore in the next chapter.

 **Now You Try!**

Creating Your Dream

Arnim envisioned himself as he wanted to be in ten years. If you tried a similar exercise, how would you envision yourself? What would you need to do to allow that dream to unfold? Jot your ideas down under the title "Creating My Dream."

Chapter 11

The Value of MOOCs and Online Learning

Adult learning is changing dramatically. Perhaps the best way to see those changes is by taking a look at a special group of people—"super-MOOCers"—who have taken a dozen, or even dozens, of MOOCs. We'll start with our old friend Arnim Rodeck, who, as it happens, is not only a master carpenter but also a master-MOOCer, with more than forty MOOCs under his belt. Since we know something about Arnim's past, it's easier to understand how super-MOOCing has brought him into the present. We'll go on to see how other super-MOOCers are making use of their learning.

> I love learning things and I read a lot. But it's hard to find good reading material that provides an introduction to the subject while simultaneously being targeted and covering important areas of that field. Luckily for me, that's what a lot of MOOCs do!
>
> —Kashyap Tumkur,
> software engineer at Verily Life Sciences

All of this is a lead-up in our journey to a *very* special place. (Hint: It has a low ceiling.)

Armin Discovers MOOCs

A decade before I met Arnim, when he was still working in high tech, his employer had been very generous with training programs and opportunities. However, as Arnim began preparing to shift to working for himself, he realized that he faced a problem. How could he keep up with learning without the kind of in-house training his employer had supplied? (Arnim's problem wasn't uncommon—it has become a major issue with the emergence of the "gig economy," where people are choosing to work more as independent contractors than full-time employees.)

More than that, at his high-tech company, Arnim had gotten in the habit of absorbing new insights by hanging around highly informative people. But there'd be no one but the cat most of the time in his woodworking studio. Also Arnim's high-tech environment had been all about change. He worried that switching to a craft as old and seemingly static as woodworking could become an intellectual dead end.

To keep this from happening, Arnim began selecting woodworking commissions that continually forced him to learn new woodworking approaches and techniques. He also initiated a routine in which he studied something new every morning for at least an hour, using library books, podcasts, and blogs.

Several years before, Arnim had learned from a Buddhist monk about the importance of setting his mind and spirit on the right path first thing each morning. The monk pointed out to Arnim how news of current events—"if it bleeds, it leads"—can instill fear and concern about issues that are irrelevant and can even be harmful to the attitudes that launched his day. (This is reminiscent of what Claudia found in relation to her depression, as discussed in Chapter 2.)

As a result of the monk's advice, Arnim stays away from news and e-mail on mornings when he's working. He tends to arise early, but he lies in bed with his eyes closed reviewing his learning and vocabulary from the day before. Next, he visualizes the *what* and *how* of the projects he is going to work on during the coming day.

Over the years, Arnim has tried to keep his mind fresh and open for hard-to-master new topics and ideas. But he noticed that the tougher the material, the more challenging it was to pick up on his own. Textbooks in subjects like philosophy or modern art sometimes seemed impenetrable. At the same time, podcasts and blogs didn't help much, because they didn't explore subjects with the kind of depth and breadth he was looking for. There were online videos, but they were often practical in nature, covering topics like how to use a table saw or camera.

What Arnim longed for were outstanding instructors with real expertise—professors who could crystallize the essence of the material and convey it in an easy-to-understand way. Arnim also wished he could actively engage with his fellow students and work with the material as he had back in college.

In 2012, Arnim chanced upon a TED talk titled "What We're Learning from Online Education," by Daphne Koller, the cofounder of a newly formed company called Coursera, which worked with universities to make some of their courses available online as MOOCs. Daphne talked about how Coursera's MOOCs were opening new horizons for learners around the world. Their MOOCs were made up of much more than just videos—they also had discussions, quizzes, and peer-graded assignments that supported learners as they grappled with the material. It sounded like the college experience Arnim was looking for.

Only part of this MOOC approach to online learning was new—many universities had done online courses for years. What *was* new was that these classes, by Coursera and others, were widely available and cheap—many were even free. There was a sort of story line to MOOCs—they had a beginning, a middle, and an end. You could go through with a cohort of fellow students, some of whom would become your buddies.

There was usually some sort of gamification—students could at least see how they progressed in the class. And there was a winning payoff in the end—a certificate from a top university like Stanford, Yale, or Princeton. These classes were also unique in how *big* they were—with tens of thousands and even hundreds of thousands of students. This was part of their allure—the sheer scale of the classes made not only for dramatic cost reductions but also gave students an opportunity to interact internationally.

Arnim was captivated by Daphne's talk, and he signed up for the first MOOC he found—Scott Page's "Model Thinking." Scott's course wasn't heavy on special effects or expensive production values. What it *was* heavy on was insight into how to use mathematical models to organize information, make forecasts, and make better decisions.

Arnim dove into MOOCs. He also spent time with the supplemental material from each MOOC—books and textbooks. He found that with MOOCs, even dense mathematical formulas and complicated philosophical ideas became more understandable. It was as if he had the professor beside him, walking him through the harder parts, doubly reinforcing what he was learning. MOOCs reminded Arnim of his old days at the university—except that he didn't have to put his life on complete or partial hold in order to attend classes.

Arnim saw that there was no limit to subjects he could explore. To his surprise, he discovered that some MOOCs are so good that even experts in the field were taking them—*as students alongside Arnim*. This provided an extraordinary opportunity for Arnim to learn from experts beyond his professors. His hour or two of MOOCing in the morning became one of the most exhilarating parts of his day.

In the last four years, the forty-plus MOOCs Arnim has taken have created a deep shift in Arnim's approach to learning. He says:

> Last year, I was with my wife in Lisbon in the famous museum of modern art. I had a hard time in it. I didn't like most of what I saw and didn't understand why it was called art. But the fact that so many peo-

> ple enjoyed what they saw made me wonder. So off I went and took a whole series of art-related MOOCs and read related books. I am far from an expert, but I have completely changed how I see art and how I appreciate it. And now my work is also changing.

Arnim likes to take MOOCs on related topics from different universities. This way, he gets a broad picture of thinking across fields, from different universities—an opportunity that's been impossible in the past for most cash-strapped college students. He realizes that he is learning *now*, in his maturity, in a way he wasn't able to before—seeing connections that he didn't have the life experience to recognize when he was a young man first in college.

Arnim's personal friendships have expanded with MOOCs. He and his wife have regular get-togethers with fellow MOOCers—local friends he has introduced to the world of MOOCs—so they can discuss

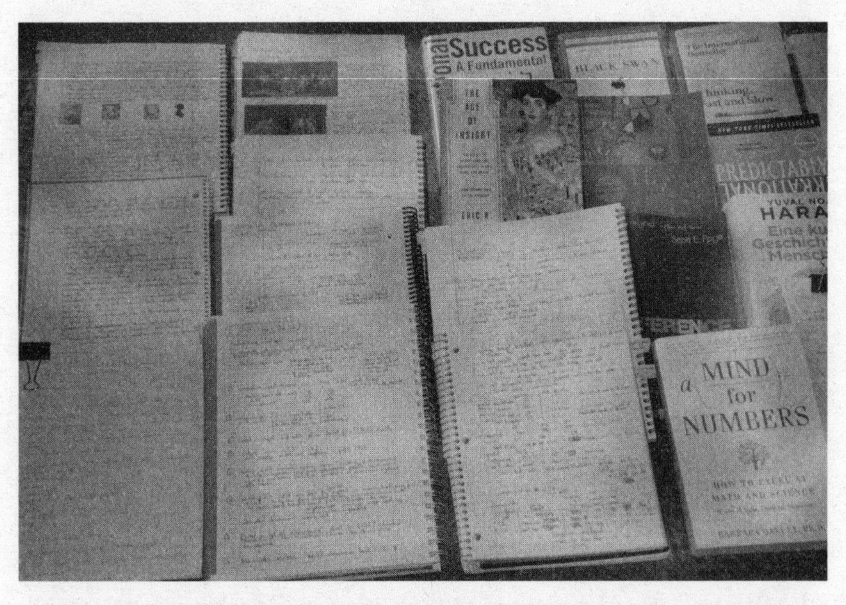

Here are some of Arnim's MOOC notes and related reading materials. MOOCs have provided a way for Arnim to continue advanced, university-level learning even as he has an active family life and is intensely involved in the day-to-day of running his business.

what they've just learned and see the material through the sometimes differing perspectives of others. As he notes, "These MOOCs have changed my life and continue to do so. I am [virtually] traveling around the world visiting the best universities. All for low cost, but of course, lots of time and energy. Yes, learning is hard. That is because by 'learning,' I mean truly changing and being able to see and think in ways I didn't before."

Super-MOOCers like Arnim and others form a new breed that gives us insight into how MOOCs can change entire approaches to both credentialing and learning.

 **Key Insight**

The Value of Opportunities to Learn

Consider learning opportunities as an important factor when making career and job decisions. How supportive is the new environment of new learning?

Super-MOOCing

Taking a dozen or more MOOCs—sometimes three or four dozen—gives a special perspective on what's available now in the world of online learning. Even the fact that super-MOOCers exist is an indication that people are finding a special sense of challenge and fulfillment with MOOCs—something akin to the allure of chess, the thrill of a poker competition, or the social feeling of quilting bees. This means it's worth it for us to take a little time to meet a bigger group of these wide-ranging, self-propelled learners.

Among this intense, online-savvy group, conventional college degrees still seem to matter: many super-MOOCers already have them. MOOCs are often being used to flesh out a new expertise—to second-skill—in a more flexible, low-cost way. Super-MOOCers are aware that many employers want self-motivated learners who keep themselves up-to-date and who are also willing to broaden their skillset.

But perhaps it's best to learn about motivation directly from super-MOOCers themselves.[1] Here we go:

"No-Pay MBA"

Laurie Pickard is stationed in Kigali, Rwanda, working for the U.S. Agency for International Development, a federal agency that manages the majority of the U.S. foreign aid budget. She holds an undergraduate degree in politics from Oberlin College, has taught public school in Philadelphia, and earned a master's degree in geography from Temple University. Laurie got her start in the field of international development as a Peace Corps volunteer in Nicaragua.

Laurie has taken around thirty MOOCs (she stopped counting after twenty) as part of her project to complete the educational equivalent of an MBA. She calls this project her "No-Pay MBA" and she has been blogging about it at www.nopaymba.com since late 2013. Laurie admits that her education was not completely free—her biggest expense has been in maintaining high-speed Internet while stationed in Central Africa. But compared with the price of an MBA degree program, MOOCs have helped Laurie save a small fortune.

Laurie Pickard "super-MOOCing" in Rwanda as she heads toward the equivalent of an MBA degree.

Laurie's work involves forming partnerships with private-sector companies to improve the lives of people in developing countries. What she especially appreciates from her MOOC learning is how she is able to directly apply it to her work in international development. She says:

My particular focus is on entrepreneurship and public-private partnerships. I've been able to develop new skills, keep my knowledge fresh, and have even been promoted since starting my MOOC education. I am able to use what I learn in real time, which is a huge benefit over full-time degree programs. Knowing that I can speak the language of business, I feel much more confident when discussing partnership opportunities with high-level executives from private companies. After covering the basics, I've developed an on-demand style in which I take MOOCs as needed to build new skills and knowledge. I'm currently taking a course on two-sided marketplace businesses (think Uber and Airbnb) in Africa. I have felt my worldview shift, I have developed a new vocabulary, and I have met fellow learners from around the world.

A Do-It-Yourself Data Science Master's Program

Twenty-three-year-old David Venturi's first internship for a dual-degree program in chemical engineering and economics left him less than enthralled with chemical engineering. A chance encounter with a friend who had interviewed with a MOOC provider led him to MOOCs. He ended up trying Udacity's CS101—an introduction to programming and computer science. (Udacity is a MOOC provider that focuses more on vocational courses for professionals, rather than general university courses, although it has recently partnered with Georgia Tech to do a master's in computer science.)

A lightbulb clicked on: programming was the subject and career David had been searching for! But how could he transition into the field? The most straightforward way David could see was to earn another bachelor's degree—this time in computer science, since he didn't have the prerequisites for a master's.

David applied and was accepted by the University of Toronto, one of the best programs in Canada. He was elated. Off he went to his new university, planning to complete their computer science program even as he was finalizing his dual degree at nearby Queen's University. He began attending classes, but soon came to a disconcerting realization—these courses were nowhere near as good as the courses he had been taking online. Another reality set in. Tuition was around ten thousand Canadian dollars a year. It would take him *three* years to graduate with a second bachelor's degree—all while earning little income but accruing a lot of debt. By the end of the second week, he dropped out of the Toronto program. He explains that the traditional university route for computer science "just didn't feel right":

> I remember sitting in the lecture hall and thinking the learning process seemed so slow and inefficient compared to my Udacity experience. I've always hated long lectures . . . I usually have to relearn all the content back at home. With MOOCs' less-passive listening and more-active doing approach and the "pause video" button, I could learn faster, more efficiently, and for a fraction of the cost.

Even as he is nearly finished with his dual-degree program, David is now in the midst of his own data science master's program using online resources. He's about 50 percent through his curriculum, which is a compilation of about thirty individual MOOCS.[2]

In addition to providing more efficient learning for many students, MOOCs offer huge cost savings over traditional college programs. David estimates a total cost of just over one thousand Canadian dollars for the program he came up with. In comparison, he says, "Paying $30,000-plus to go back to school seemed irresponsible."

The Good and the Bad of Do-It-Yourself Degree Programs

Here are David's own reflections on how his do-it-yourself program has worked out for him.

GOOD

- **Because of MOOCs, I finally found a career path I'm excited about!**
- I'm **saving tens of thousands of dollars**, not including the opportunity cost of getting into the workplace faster.
- **No forced extra electives.** I'm learning what I want to learn, faster, more efficiently—and at a lower cost. I handpick my courses—an important consideration, because I already took plenty from my original undergraduate degree.
- I can **fast-track my studies** without trudging along with a strict four-month semester time line.
- I'm **learning where and when I want**. It's awesome to have the freedom to make my own schedule and choose where my "classroom" will be. All I need is a laptop and headphones. I also feel less pressure than I did when studying for my chemical engineering degree, though I'm working the same amount of hours per week as I did back then. This is probably due to a combination of the increased content absorption efficiency and no hard deadlines.
- I **connect with people all over the world** through Twitter, Slack, LinkedIn (where Udacity has a Nanodegree alumni group), and my website. Next week, I'm having a Skype chat with a fellow MOOC graduate from India. That's pretty cool.
- I'm **helping and inspiring others**—which inspires me. A number of people (friends and strangers) have reached out and told me they were inspired by my path and asked me to guide them regarding their online learning goals. It is very fulfilling, both to hear they were inspired and to enable others to move forward.

- I'm **raising awareness**. Many people still aren't aware that incredible resources like Udacity, Coursera, and edX exist. I feel very lucky that the timing has worked out for me in regard to discovering MOOCs. I want people to be aware so they don't need a stroke of luck to seize a potentially life-changing source of education.

NOT SO GOOD

- **Work-life balance decisions are trickier.** A self-paced program requires self-discipline. Turning down social stuff to stay on schedule is way more difficult without professor-imposed, grade-threatening deadlines. When taking a MOOC, you are presented with more decisions to make regarding time allocation for family, friends, physical activity, entertainment, school, networking, and sleep.
- **I lack the level of classmate interaction of a traditional university.** The lack of face-to-face interaction is mitigated by the ability to connect with people all over the world. But it'd be more awesome if we could also hang out with nearby MOOC-mates. Services like Meetup and Udacity Connect are trying to fix this, but they aren't that well developed yet.
- **Flexible deadlines make for doubt and guilt.** It's never about whether I am learning the content, it's: *Am I going fast enough? Did I work enough today? Holy crap, going to the gym and making meals take four hours out of my day!* I never felt unsatisfied with my progress in the university, where everyone had the same deadlines. When you're making your own schedule and creating a unique mix of courses, you're also creating your own benchmarks for what you want to achieve.

With MOOCs, Browsing Is Welcome and Failure *Is* an Option

Pat Bowden is a retiree in Queensland, Australia, who started taking MOOCs because she saw it as a way to pursue interests she had been unable to follow earlier in her life. After failing her very first MOOC, on astronomy (it had been forty years since high school physics), she's gone on to successfully complete seventy-one—and to fail another twelve.

Former bank officer Pat says, "MOOCs have opened up a new world for me. I envisaged a retirement doing handcrafts and pottering around the garden. Instead, despite a rocky start to my MOOC studies, I have learned a lot in the last few years and now have the confidence to start a writing career."

It's worth noting what it really means to "fail" a MOOC. It's a different kind of failure from a traditional classroom. For one thing, the stakes are lower—MOOC grades are not going to go on a college transcript. For another, there are opportunities to try again—if you don't pass a MOOC, it's usually offered again within another few months.

Laurie Pickard, the "No-Pay MBA" foreign aid worker in Rwanda, observes that it's impossible to truly fail a MOOC in the traditional sense, since MOOCs generally mean learning at your own pace and exploring the limits of your current capabilities. Another nice feature of MOOCs is that it's easy to back away from a MOOC and take another MOOC or two to get the background knowledge you may need to be successful in the first MOOC.

Education technology entrepreneur and super-MOOCer Yoni Dayan (we'll learn more about him further on) notes that many learners are only interested in parts of a MOOC, or in quickly browsing through the material and watching specific videos. They may not complete a MOOC, but they've used the MOOC to help them succeed in their own learning goals. In this sense, MOOCs are a lot like textbooks. Students are made to purchase $250 textbooks, of which they may use less than half the material. Yet no one says that textbooks aren't worthwhile because of their low "completion" rates.

Self-Improvement That Also Enhances Business-Related Skills

Cristian Artoni, an operations manager and analyst on the staff of the chief operating officer for a major Italian transportation company, has taken nearly fifty MOOCs. An avid learner, he also reads at least a dozen books a year, always related to the themes of the MOOCs he is following.

On his LinkedIn profile, the list of MOOCs Cristian has taken reveals an extraordinarily broad range of interests—including ancient philosophy, management, spreadsheets, public speaking, negotiation, and, of course, "Learning How to Learn." This may seem a wild hodgepodge, but Cristian uses a precise logic to approach his MOOC-taking. In essence, he takes theoretical and practical concepts from the courses, which not only improve him as a person, but also allow him to function more effectively at his work.

A cornerstone of Cristian's studies has been learning *how to learn effectively.* With this skill, he feels he's been better able to access a range of other skills. Since Cristian plays the role of mentor, coach, and trainer within his company, he's found that knowing how people learn has been particularly worthwhile. Also useful have been MOOCs in leadership, communication, and negotiation, which have enhanced Cristian's ability to find new ideas, communicate them, and then persuade people to act on them.

Cristian also values critical thinking. He notes: "Philosophy is the foundation of this skill, while logic is its tool." Related to this are problem solving and time management, which are key requirements for Cristian's everyday work.

Cristian became a senior mentor for the MOOC "Learning How to Learn" and then took charge of the Italian version of the course. Under

I have recently begun to experiment with online learning. It's both interesting and relaxing. I choose when I want to take courses. I can also replay videos until I understand the key ideas—I can't do that with my teachers in a regular class. Online is the best way I've found to learn new skills.

—Sanou Do Edmond,
third-year statistics student from Burkina Faso

Cristian's leadership, gathering fifty volunteers together and translating the equivalent of an encyclopedia happened virtually overnight. His organizational training from MOOCs, as well as his own natural acumen, have allowed us to improve the management of a course so large that it dwarfs any previous approach to teaching.

Fine-Tuning and Expanding Technical Skills

As a program evaluator and database administrator in the human services nonprofit sector, Jason Cherry has found that MOOCs enhance his job performance. Most of Jason's colleagues are social workers, which has meant that it has been tough for him to fine-tune his technical skills with fellow high-tech types at work. At first, Jason took MOOCs to improve his analytics skills and learn web development and more programming. Once he started, he couldn't stop—to date, he has completed roughly thirty-five MOOCs. Jason says, "One thing I really like about MOOCs is the flexibility—still having deadlines, but also allowing me to go as rapidly as I'd like. It's easy enough for me to blast through a week's content in one class in an afternoon." Right now, Jason is supporting his work's development department with his MOOC-obtained insights in predictive modeling, which is well beyond anything they had before.

Reinventing Yourself

Brian Brookshire was the smart kid who blew the test curves in Des Moines, Iowa, while growing up. But although he had plenty of friends and girlfriends, and was even in a fraternity, social relationships never quite clicked for him. He says, "I just didn't have a good algorithm for social interactions. People would ask me why I was so quiet and I would reply, 'Because I don't know what to say.'"

He graduated from Stanford with a bachelor of arts in Japanese, and a few months later, he found himself in Korea studying in an intense, yearlong Korean language program. One evening, he was killing time online when he ran across a quiz that asked, "Can you tell when she's ready to be kissed?" Intrigued, he clicked on it. At the end an ad popped up for a video course on dating. On a lark, expecting nothing of real value, Brian enrolled. But a whirlwind tour of ideas in evolutionary psychology and general self-help, and adaptations of techniques from sales and marketing, soon convinced him of an idea that he found revolutionary—social skills were indeed learnable.

Brian says:

> I quickly discovered that there was an entire category of dating advice for men seminars being promoted on the web, and a host of web forums where the material was being discussed. I tried it. I tried it all. Surprisingly, a lot of the material was not even really about dating per se. There was a lot of emphasis on perspective-taking and being able to understand where others are coming from. Realizing that other folks are walking through life experiencing the same worries, concerns, and hopes as you. As one speaker put it, "The more personal an issue seems, the more universal it is."
>
> It worked. Over a six-month period I met and went on dates with approximately sixty different women—one of whom I eventually married. The effects spilled over into all other facets of my life. I began to feel at ease in social interactions in a way that I had never felt before. Meeting new people became a genuinely exciting experience.

This self-reinvention through online learning was a number of years ago—but it taught Brian a lot about how learning could make extraordinary changes in his life—much broader than he'd ever thought possible from the usual academic approaches. He has set a powerlifting national record, has appeared in fashion magazines, speaks Japanese at the highest levels of proficiency, and is fluent as well in Korean. Not bad for an ordinary "meat and potatoes" guy from Des Moines.

These previous self-reinventions have led to Brian's most recent learning adventure. Wanting to understand more about the microbiome, he took a MOOC. Before he knew it, Brian was fifteen biology courses deep and enjoying himself more than he had in years. A question he always likes to ask himself is, "How can I take what I'm doing to the next level?" He thought of setting out to complete a PhD in biology. But going back to undergraduate school for three years to complete the requirements to apply to a doctoral program was unappealing. So, inspired by an online computer science program he'd seen, Brian put together an undergraduate biology program insofar as possible with the equivalent MOOCs—he's blogged his efforts.[3] Whether or not Brian gets his doctorate remains to be seen. But his learning path, along with his preexisting business and language skills, is meanwhile positioning him to be knowledgeable about a nascent market in Asia that melds both biology and business.

Working Through Disabilities

Super-MOOCer "Hans Lefebvre" became a quadriplegic after an unfortunate fall at age eleven.[4] He types on the keyboard with a stick in his mouth or by using voice recognition. Hans has earned a master's in astrophysics, and he would like to earn a second master's in computer science. But he doesn't have the previous coursework to allow him to enter the program without starting all over with a second bachelor's degree.

As it turns out, however, the university has another path for getting in their master's program—passing their equivalence examinations. To that end, Hans has taken more than fifty computer science MOOCs, in which he has been a top student—he has even become a mentor in Princeton's algorithms MOOC. Hans still has a few more courses to take before he feels he is ready, but the availability of advanced online coursework through MOOCs has made a way for this gifted learner to have dreams for the future. Hans's long-term goal is to get a job in university research. This is not an impossible dream—he lives in one of the most accessible cities in Europe for those with physical disabilities.

Hans says, "I love learning, so I'm having lots of fun. The more I learn, the more I realize how much I'm lacking skills, but that just motivates me to keep learning."

Social Mindshifting: Using MOOCs to Develop New Social Networks

French-Israeli entrepreneur Yoni Dayan, a thirty-four-year-old graduate in international affairs from the Sorbonne, has always been fascinated by start-ups. When he was only eighteen, he cofounded a company that reviewed video games—since then, he has been drawn to the idea of creating businesses that help others.

A benefit, but simultaneously a challenge, for Yoni has been the "natural" social circles that surround him. As a college student more than a decade before, he formed professional connections in the international affairs community. But for an entrepreneur like Yoni, those types of connections weren't enough. He needed entrepreneurially oriented social networks to allow him to build his business-making acumen.

MOOCs formed the perfect networking opportunity. Yoni has taken several dozen, many in business and entrepreneurship, and others in related areas such as programming, creativity, and design. Yoni says, "By struggling together in submitting the team assignments on time, by collaborating through hangouts, and by sharing our stories, my virtual acquaintances have become friends and associates." In the past few years, the succession of small wins and continual support from the like-minded individuals Yoni has met in online courses have given him the confidence to fully embrace his inner entrepreneur. Among many other projects, he is now in the process of leveraging the insight and networks he has developed from super-MOOCing to create a new start-up on valuing the informal ways we acquire knowledge and skills.

Being a Generalist

Super-MOOCer Paul Hundal, a fifty-nine-year-old lawyer from Vancouver, recently finished his hundredth MOOC through edX. In an age when having specific programming or business skills is valuable, it's easy to forget that society also needs generalists. And, as Paul points out, being a lawyer is about being a generalist. Rather than having specialized knowledge on one topic like a scientist, a lawyer has to analyze every case as a new fact pattern with different and sometimes hidden issues. The more generalized knowledge he has, the better Paul can analyze a situation.

For the past two decades Paul has been on the board of directors of one of Canada's oldest environmental groups, the Society Promoting Environmental Conservation, where he has led campaigns for water- and air-quality protection, the protection of old-growth forests, wildlife habitat conservation, and waste reduction. These campaigns required Paul to develop generalized knowledge from many disciplines so he could be effective at advocating for conservation.

Paul's approach has always been to get the facts and science right. He recalls, "Twenty-five years ago, when I needed to quickly learn something in a specialized field, I used to call the local university and speak to an academic expert. At that time professors were surprisingly open to talk to someone on a cold call, especially when they knew why I was asking. Over the years, however, that became much more difficult. People just are not as accessible through cold calls anymore."

Lots of information is now available on the Internet, but learning to sort the good from the bad is key—twenty sources can cite the same false statement, which came from the same original bad source. Paul's been forced to become more of a do-it-yourselfer in finding the knowledge he needs to do his job properly. He says, "When I first heard of edX, free online courses taught by the best professors in the world, I immediately was convinced of the value of it. I could study almost any topic, benefiting from the best scholarship in the world, quickly and easily from home. After one hundred MOOCs it has been a wonderful experience in broadening my knowledge in the most efficient way so I can share that knowledge with others. It makes me a better lawyer and a better environmentalist advocating for conservation and a better world."

Key Mindshift

With New Forms of Learning, You're the Driver

Remember that the new forms of learning allow *you* to take charge. MOOCs are an important new resource for achieving your learning goals, whether they demand technical skills, soft skills, or even skills about learning itself!

Changing Your Brain: Online Learning Makes It Easier

Jonathan Kroll is an entrepreneur who loves learning languages. And that's a bit of an understatement. Back in college at UC Santa Barbara, he majored in French and Spanish and minored in Portuguese (but only because he wasn't allowed to triple major). He had also studied Latin, Italian, and Catalan.

Jonathan's aptitude for languages was not matched by an aptitude for math. In fact, Jonathan was terrible at math. However, this wasn't a big deal to him, because he was planning for a career in foreign service.

Shortly before graduation, though, Jonathan found himself drawn to the potential of the then newly exploding Internet. There were entrepreneurial opportunities galore, and Jonathan threw himself headlong into the competition with others who were forming new companies and services like Facebook, YouTube, and Gmail.

By day, Jonathan would go to class to learn and bone up on the sort of languages he'd always done so well in, and at night, he'd come home to research, learn, and experiment with programming languages. He was surprised to find that he had transferable skills—the grammar, syntax, and semantics of the languages he had been studying for so many years had primed his brain to easily digest and understand the rules governing computer languages. (Yet again, we see that "irrelevant" expertise from the past can provide a surprising advantage in a new career.)

Eventually, in hopes of giving himself a better understanding of the entrepreneurial world he was embracing, Jonathan set his sights on business school. He began preparing for the GMAT (Graduate Management Admission Test). The GMAT is math-intensive, so Jonathan knew it would be tough on him. There are no calculators allowed—Jonathan would have to do every single calculation by hand during the test. He would have less than two minutes per question, so every second mattered. However, Jonathan, at twenty-nine, could barely perform simple multiplication or division without the aid of a calculator—much less factor a polynomial or calculate the number of permutations of *n* distinct objects in a circular arrangement.

He took the test and got the results. It was pretty much akin to having been run over by a semitruck. He wasn't just bad at math—apparently, a first grader could have done better.

He picked himself up, dusted himself off, and started over with his studies. Understanding now how bereft he was of math fundamentals, he decided to start his reviewing process with elementary school math. He worked with tutors and test-prep experts, and studied for hours on his own. Little by little, he mastered each concept.

Over two years, he sat for the four-hour GMAT exam on *six* different occasions. On top of that, he sat for the GRE (Graduate Record Examination) four times. In the end, his scores blew away those of the vast majority of Americans. What was really important was how, through those two years of study, Jonathan gained a completely different outlook on his ability for math. He really *could* do it after all.

Jonathan knew there had to be a better way for people to study for the quantitative sections (the parts with all the math-related questions) of exams like the GMAT and the GRE. He and his former tutors realized they had an opportunity at hand, in the form of an already-existing company named Target Test Prep. It was a company with not fully realized potential. Its assets were its comprehensive curriculum and thousands of proprietary practice questions. But its software was outdated, its branding was lackluster, and its market penetration was poor. Jonathan saw the possibilities and decided to delay his plans for business school to instead join Target Test Prep as chief technology officer. Within weeks, a plan had been crafted to rebuild its GMAT software from scratch—and a month later, a ten-person development team had been assembled and was ready to get to work.

MOOCs were just becoming more widely known, and out of curiosity Jonathan had already taken a few. He was surprised to find that this new knowledge was immediately useful.

For one thing, panic is a big factor on high-stakes tests. Jonathan had been there himself—the pressure turns your mind blank, you feel almost frozen, tunnel vision sets in, the clock runs, and the stress makes applying even your most comfortable concept nearly impossible. The MIT MOOC "Design and Development of Educational Technology" taught Jonathan about "active learning" and how noncognitive skills can be as important as content knowledge when preparing for a high-pressure, high-stress environment such as the GMAT. This allowed Jonathan to develop insights and ideas for features that he brought to the table during planning discussions.

In a related vein, it turns out that the GMAT contains so many concepts that it can be overwhelming for students to know how to organize their studies. From the MOOC "Learning How to Learn," Jonathan had learned about the concept of chunking. (If you'll remember from Chapter 3, chunking involves building small chunks of knowledge using day-by-day practice and repetition. It's the foundation of expertise in any subject.)

Knowledge of chunking led Jonathan to push to have Target Test Prep's content organized online so that each unit was small enough to "chunk." His team then designed a system for students to practice questions related to each specific chunk. For example, they took the concept of "exponents and roots" and made it its own pedagogical unit—an "über-chunk." This large chunk was then further subdivided into fifty or so smaller chunks, each with its own corresponding body of practice questions. All of this was built into the very deepest structure of the software. Such an approach may seem common sense, but it is not something done to the same degree by other test-prep companies, where, for example, everything related to exponents might be swept into a big "arithmetic" category.

Jonathan's experience as a "bad" math student was surprisingly valuable. For one thing, he knew exactly where students would struggle. As Singapore's Adam Khoo would say, Jonathan turned a problem—his inability to do math—into an opportunity.

Since its relaunch, Target Test Prep has been featured in top magazines and has formed partnerships with leading universities and organizations.[5] They have helped thousands of students earn impressive scores on the GMAT, GRE, and MCAT—and perhaps as important, the company has allowed people to improve the critical-thinking and ana-

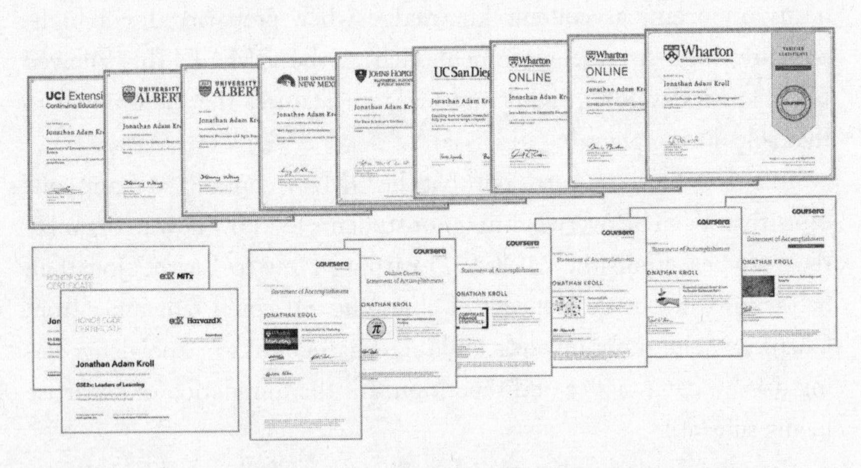

Jonathan Kroll's MOOC certificate collection.
MOOCs have allowed Jonathan to bring new ideas from neuroscientific research directly into the creation of a new, useful, and popular product.

lytical skills that are in strong demand today. It's not just teaching to the test—it's teaching vitally important skills.

As for Jonathan? Although his math studies were done before MOOCs came along, he now understands the power of learning in making pivotal life changes. He has become a super-MOOCer, completing eighteen MOOCs to date. He is always learning something new, either for professional improvement or out of simple curiosity.

Key Mindshift

Online Learning Provides a Great Path for Renewal!

It can be shocking to realize how high school skills have atrophied or were barely there in the first place. Online learning provides a great way to refresh old learning, bone up on the skills you need for a critical test, or simply gain foundational skills.

Super-MOOCer Ronny De Winter's Tips for Getting the Most Out of MOOCs

Ronny De Winter is a freelance software engineer from Belgium who has completed fifty MOOCs. Here are his tips for top performance:

- Define what you would like to learn in the next two to three years.[6]
- Find the MOOCs and other learning means that best fit your needs—Class-Central.com is very useful here.
- Before enrolling in any MOOC, carefully investigate the course outline, prerequisites, syllabus, and suggested weekly workload.
- Schedule time each week. To play it safe, it's a good idea to arrange to have twice the recommended time available for your learning.
- Some people like to listen to videos at anywhere between 1.2 to 2.0 times normal speed. More advanced MOOCers sometimes employ

"fast-MOOCing." In this approach, you skim through the syllabus and slides before watching videos. Then you watch the videos at up to 2x speed.

- See how things go during the first week. If you're not getting a lot out of a MOOC, stop taking it.
- Don't enroll in too many courses at the same time. It's better to study a few subjects deeply than many courses superficially. Most courses are repeated, so you can usually take a course later if it doesn't fit into your schedule.
- Use the discussion forum to enhance your learning and to get questions answered—but be aware that this can be time-consuming.
- Any time you enroll in a brand-new course, you may find it a bit buggy, so if this bothers you, wait until a later offering. But first-time courses can still be fun, so don't discount them altogether.

Can We Learn Too Much?

Sometimes MOOCs are valuable because they cause us to step back and look at what we really want to learn, and why. Ana Belén Sánchez Prieto, professor of medieval manuscript studies and archives management in Madrid, gives interesting insight about this. Ana was originally a total skeptic about the value of online learning. But when her university started a pilot program for an online master's degree, she volunteered to teach a class, mainly because she figured that teaching via the web would allow her to spend more time with her husband, who worked abroad. After all—if she taught online, she could teach from anywhere.

Since Ana was going to create an online class, she wanted to try one. She took the MOOC "Archaeology's Little Dirty Secrets," by Sue Alcock, a classical archaeologist at Brown University. Ana loved it—it gave her a lot of ideas for her own online class. But oddly enough, it also showed Ana that, just because she knew her subject, it didn't mean that she knew the most effective way to teach it. With that in mind, Ana enrolled in another MOOC, "Foundations of Teaching for Learning." She thought that was a good class, too.

Next, Ana realized that there was a whole cluster of education-related MOOCs, a "specialization," from Coursera. *Hey, a specialization would look good on my résumé*, she realized—though she was a full professor with tenure who loved her job and had no intention of moving.

Ana ended up taking every MOOC she could find about education. However, the MOOC "Teaching Character and Creating Positive Classrooms," by Dave Levin of the Relay Graduate School of Education, formed a special turning point. By the time Ana got to the second half of the course, her husband was asking, "Ana, what's happened to you?" Ana says, "I do not know if the MOOC helped me to become a better teacher, but it definitely helped me to become a better human being." She feels that the MOOC gave her a deeper understanding of others and a willingness to forgive their weaknesses.

Then Ana began to discover MOOCs on subjects she had always wondered about, but had never had the chance to learn, such as computers. Dr. Charles Severance's classes on the Internet and Python were fantastic eye-openers for her. (Side note: Most of the MOOC-making and MOOC-watching world loves "Dr. Chuck.")

Then Ana began taking classes on HTML and other tools for web development.

Then she began working her way through math with Khan Academy.

Ana was so excited by her ability to learn—and get credentials for that learning—that she got carried away, taking MOOCs on practically everything. "It caused real stress, because I also have my own classes. My social life began vanishing. Finally, I had to face it—I was addicted to MOOCs. And what is worse, I often wasn't really learning, because I was focused more on finishing and getting the certificate." By that time, she'd gotten around fifty MOOC certificates and ninety-one Khan Academy badges.

Ana stopped cold turkey. She had finally realized that many things can be interesting to learn about, but she had to choose.

Professor Ana Belén Sánchez Prieto, on the right, as she prepares to present for her MOOC on medieval manuscripts.

A few months after her cold turkey withdrawal from MOOCs, Ana went back to MOOCing. But this time, she took a more balanced approach. She's started "sitting in on" a MOOC on game design so she can apply gaming techniques to improve her classes. She's planning to repeat the gaming class again before formally enrolling, so she really learns the material. Ana's objective now is simply to learn—but not to be overwhelmed.

By the time you read this, Ana's MOOC "Deciphering Secrets: The Illuminated Manuscripts of Medieval Europe" will have launched.

Key Mindshift

Balance

Life holds many—sometimes too many—learning opportunities. If you are just getting started in the world of MOOCs, note that they can be addictive. If a subject interests you, you can audit a course to browse the material when and where you want, without the pressure of assignments and deadlines. Certificates can be great for motivating yourself. But use common sense to balance learning and certificates with professional obligations and family life.

Why MOOCs and Other Online Learning Courses Matter

Perhaps you're wondering why I've been putting so much emphasis on MOOCs and online coursework in this chapter, as opposed to plain old television or videos. The catch is that television and videos often contain passive, "watch only" sorts of materials (with a few important exceptions we'll get to soon). This means that television and video can form a great start for learning, but it's often still not enough. Many people need a bit of a boost to truly get their brains wrapped around the material. Well-designed MOOCs provide that boost—they help bring the material to life through active learning—the kind of learning that leads to more profound physical changes in the brain. Active learning, remember, is what Jonathan Kroll found to be so valuable when designing an outstanding test-prep system for the GMAT. Such fundamental neural shifts have an impact, not only on mental flexibility, but on your long-term health and longevity.

Here's what I mean. In passive learning, you might watch a TV show and discover there's a musical instrument called an oboe. In active learning, *you become able to play the oboe yourself.* Active learning is exceptionally powerful—it's the kind of learning that allows you to make logical arguments, formulate good questions, problem-solve, expertly kick a soccer ball, speak a foreign language, play that musical instrument, or simply be more creative with what you're learning.[7] (Ever wonder why we have the "Now You Try!" sections in this book?)

MOOCs, especially well-designed MOOCs, provide plenty of structured opportunities for active learning through their tests, homework assignments, projects, and discussion forums. Even if all you do is take a few quizzes while breezing through a MOOC's videos, you'll see that those quizzes allow you to understand the material in a fresh way. And of course, quizzes reinforce your new knowledge and allow you to check to see whether you really understood the material. (Naturally, deep quizzes involving fundamental concepts are much better than shallow quizzes with superficial, "memorize this" type answers.[8] But deep quizzes can be much more difficult to create.) All this said, sometimes it's just not necessary to go overboard with active learning—especially when all you need is a bit of an overview of an area. Again, this is why MOOCs can be so handy—you can pop in and just get what you want or need.

The great thing for learners is that MOOCs compete with one another. All you have to do is go to an outfit like Class-Central.com, which analyzes and compares MOOCs. You can pull up their ratings to find the best MOOC on negotiation, public speaking, organic chemistry, or what have you. It can also be fun to read the reviews, which sometimes seem like they're straight out of Rotten Tomatoes.

There are challenges, however. At present, many students have trouble motivating themselves to complete the active parts of MOOCs. This is why coding programs like Dev Bootcamp, which has a large face-to-face component, can be so worthwhile despite their high cost. Another problem is that MOOCs provide an education that, in general, won't supply credits toward a college degree. (New technology that applies airport-style facial recognition technologies to proctor MOOCs may spark big changes here.) And yet another issue is that many current MOOCs are too conventional in their structure and passive in their pedagogy. Professors drone on through long lectures that have been sliced into parts to make the videos seem shorter, while the only "active" parts of the course are cursory quizzes with superficial projects. None of this is enough for you to truly wrap your mind around a topic and embrace learning by doing and practicing.

Super-MOOCer Jonathan Kroll has observed that we are moving toward an à la carte skills credentialing model where education becomes more of a salad bar than a sit-down, table-service restaurant. There are smart online education companies that are beginning to make sense of the multiplicity of ways of learning. If you go to the online company Degreed, for example, you'll see ways to input your own learning from hundreds of different platforms like Khan Academy, Coursera, and

Udacity, as well as from books, TED talks, articles, and university coursework and degrees. Degreed's motto is "A million ways to learn—one place to discover, track, and measure all of it."

In any case, now that you have an introduction to MOOCs and how they fit more generally with the world of online learning, we'll next discover what it's like to do online learning from the other side of the camera. We're finally going to that low-ceiling place I told you about.

Chapter 12

MOOC-Making

A View from the Trenches

I'M A PRETTY straightforward, old-fashioned midwestern engineer—the kind of person who's happy to be asked out to lunch with friends at McDonald's. So it was kind of a shocker for me when I was invited to speak at Harvard about "Learning How to Learn," the MOOC I created with Salk Institute neuro-ninja Terry Sejnowski. I was even more surprised to arrive in Cambridge and see the room packed with Harvard and MIT folks, all eager to learn the "secret sauce" behind the making of our MOOC.

Eventually, I understood—at least in part—the reason for their curiosity. "Learning How to Learn" was a labor of love made for less than $5,000. Yet it had on the order of the same number of students as *all* of Harvard's dozens of MOOCs put together, made for millions of dollars with hundreds of people.[1]

Strangely enough (though I didn't share this with the audience), one of the motivators behind my making of this MOOC was the worst professor I ever had in college—call him Professor Clueless. One day, while he was at the blackboard, puzzling over some relatively simple equation he'd screwed up, the students started talking about a TV show. He whirled around to face the class, puffed his chest out, and announced: "I never watch television."

I was in my thirties then and rarely watched television myself. But because this awful professor had sneered at it, the only thing I could think of was, *I'd better start watching some TV!*

So I did. I didn't watch a lot—just a couple of hours a week. But over the past two decades, that little bit of TV-watching has given me a real appreciation for the power of video and visual imagery for conveying information. As an author, I might write a book like *A Mind for Numbers* about how to learn math or anything else. But by watching television, and talking to people who watched television and other videos online, I began to realize something important. The people who most needed the information in any book about learning would never read those kinds—or any kind—of books. *Instead, those people watch video.*

And there was nothing wrong with that. Remember how I mentioned in the last chapter that television and video don't *always* involve passive learning? Video can not only provide the groundwork for active learning (mimic me and you too can unclog your toilet!), it can serve as a motivator by fantastic guides to explore everything from ancient Greek mythology to string theory. When a video is done well, it's enjoyable—even when it's teaching difficult subjects like calculus. Integrated with the kinds of active learning support materials that MOOCs can supply, video can have a great impact on learning. Good MOOCs may not necessarily make learning *easy*, but they can help motivate you through the material, and help make it *stick*.

In the last chapter, I'd also mentioned that we were going someplace special. Well, we have finally arrived. It's the family room in our basement—the low-ceilinged video studio, where much of the MOOC "Learning How to Learn" was born. There's a photo of it on the next page. I think you'll find it worthwhile, discovering what happened in that basement. It will provide clues about what to look for when you're seeking quality online or in-person learning. My hope is also that you'll gain a better sense of the future of learning.

The upper image shows the raw footage of me in the basement. The black at the top of the photo shows a bit of the hood of the teleprompter. Those black hooded "umbrellas" on the left and the right are two studio lights. (And yes, on the far edges, you can see the fireplace and the blinds.) The lower image is the final composite image of me walking between two metaphors for focused *and* diffuse *thinking. When I was being videotaped, I was acting like the weatherperson in front of that green backdrop, so I had to imagine how it would ultimately look when I composited the footage of me with that of the PowerPoint animations I "walked" among. (Yes, a lot of the moving background imagery in "Learning How to Learn" was simple PowerPoint caught with a screen capture program.)*

Online Learning: How the Sausages Are Made

Getting Traction

Once Terry Sejnowski and I decided to do the MOOC "Learning How to Learn," we realized it wouldn't be easy. Unlike most MOOC-makers, we didn't have a big grant or some sort of solid institutional backing. But we did have one thing going for us: Terry was a professor at UC San Diego, which was plugged in to the online provider Coursera.

After I scoped out options for getting our course made, it was clear that the only real choice was to buy a camera, put together a little home studio, and do the bulk of the videotaping there. So that's what I did.

Of course, there was a problem with this approach. I had zero previous video-making or video-editing experience. I could just barely press the right button on a camera—that is, if somebody pointed it out to me. I remember, just three years ago, upon seeing a picture of someone's office video studio, thinking to myself, *Wow—no way I'd have the expertise to pull together something like that!*

To create the basement studio, I Googled "how to create a green screen studio" and "how to set up studio lighting." I watched YouTube videos on video editing and then tried things out for myself. In fact, being able to watch and then try things out myself is what formed the active part of online learning that made it all come together for me. (If you'd like, take a little journey through the endnote here for a listing of hard-earned insights.[2] A good MOOC on MOOC-making could have saved me so much trouble!)

In the "green screen" approach, you videotape yourself in front of a green backdrop—even a simple green cloth will do. Computer wizardry during the later editing process can then replace that green backdrop with whatever you'd like—for example, the "pinball" metaphors in the illustration that started this chapter. I chose a green screen approach because it would provide a lot of flexibility to allow my image to be moved around on the screen and to add cool effects—only later did I discover that green screen is considered a more advanced video technique.

You might think that learning videography was easier for me, as an engineer, than it might be for you. But the reality is, making professional-looking and -sounding educational videos—even with a "sophisticated" green screen—is not that tough for *anyone* nowadays. I won't lie to you—as with any new venture, there were moments of frustration. But whenever I'd really get stuck, I'd just ask some local high school kid for help.

Terry shot footage of his part of the MOOC in San Diego and sent it my way. I edited in his insights, which help provide the neuroscientific backbone of the course. One of the many great things about Terry is that he's not only a legendary neuroscientist, but he shows how to use neuroscientific research in practical ways to improve our lives.

Terrence Sejnowski, co-instructor of "Learning How to Learn," practices what he preaches in the course about the importance of exercise. When I visited him in California, I asked where he exercises. Next thing I know, Terry's scrambling like a mountain goat down a 400-foot cliff and running for miles on the beach.

My heroic hubby, Philip Oakley, was the man behind the camera, also running the teleprompter and handling audio. In addition, he did some of the initial rough editing. Oh, and then there was the psychological support he provided. I'd flub a take for the fourth time and rip off my microphone, melodramatically announcing, "I just can't do this!" He'd listen, then calmly tell me to get my act together and get back to work. Our son-in-law did some of the cool metaphorical imagery—surfing zombies, metabolic vampires, an octopus of attention. Our two daughters kindly "volunteered" for bit parts like backing a car up into a ditch or looking dorky in a set of oversized headphones. This kept production costs down.

The use of family, um, "actresses" would also lead to later surprises. Our older daughter, for example, a medical school student at the time, was taken aback when her professor—a distinguished specialist—suddenly stopped the class, pointing directly at her: "Wait—you're in the MOOC!"

My hubby Phil, safely behind the teleprompter as I have a diva moment.

As I worked on the MOOC, preregistrations began to add up—ten thousand, thirty thousand, eighty thousand. MOOCs don't usually see anything like this kind of interest so early on. This was scary, especially because we weren't doing anything unusual to promote the course. I didn't have time to promote the course.

Partway through production, I made the mistake of reaching out to a professor with a popular MOOC.

"Any tips you might be willing to share?" I asked.

"Why don't you talk to my producer?" he answered.

"Okay," I said, thinking, *Holy crap, this guy's got a producer!* I didn't have a penny for staff.

So I talked to the producer. She said, "Get ready to go with no sleep for six months, because trying to get the twenty people on the production team all synced together is crazy-making."

I thought, *Twenty people! Production team!*

I started to panic. I was working away, scripting, shooting, editing—fourteen, sixteen hours a day.

At that time, even inside academia, few people had even heard the word "MOOC," so I couldn't easily explain what I was doing. My super-competent editor at Tarcher/Penguin, Joanna Ng, called to urge that I do the typical author thing and write op-eds to promote my soon-to-be-published book, *A Mind for Numbers*—the book that "Learning How to Learn" is based on. I told Joanna, "I'm, um, a little preoccupied—I'm in my basement making a MOOC."

There was this long pause. Joanna was hyperpolite, like people get when they're not sure someone's quite all there. "What's a 'MOOC'?" she finally asked.

The Benefits of a Fresh Perspective

For me, the biggest drawback and—simultaneously—the biggest advantage of putting together the bulk of the MOOC myself was learning to edit video. As it turns out, video editing is one of the most time-consuming, most expensive, and—as I learned—most critical aspects of production, because editing goes right to the heart of what sparks our attention. And attention is essential in learning.

It's worth noting that in television- and moviemaking, production and editing revolves around creating compelling sounds, visuals, and stories, in order to get people to *pay attention to what's on-screen*. In academia, on the other hand, the focus is on *creating the prescribed number of hours of educational content*—important for accreditation purposes. Sadly, the academic tradition of "just doing time" has extended to the production and editing of a surprising number of today's MOOCs. Though there may be high production values, those alone don't make a MOOC watchable or worth learning from. To understand what does, let's visit MOOC-making together—I'll be your guide to an insider's perspective on MOOC-making.

Key Mindshift

The Value of Fresh Perspectives

There can sometimes be great value in striking off on your own and not following traditional approaches. Even though it may be intimidating, look for opportunities to bring your own unique insights and fresh ways of doing things into your work or your hobby.

Dhawal Shah's Story: An Eye for Opportunity and a Willingness to Learn

Learning on the fly when the right opportunity might arise is part of the key to success. For example, take Dhawal Shah, the founder of Class-Central.com, a website where people can find the highest-rated MOOCs on any topic they might choose—much like Amazon provides a rating system for books. Dhawal says:

> Class Central was something I built for myself over a lonely Thanksgiving weekend in Dallas where I was working as a software engineer. All my friends had gone to visit their families, so I had nothing to do. But I was really excited about free online

courses—MOOCs—from Stanford that were being announced. So I built a simple one-page site to keep track of these courses. I shared the link to Class-Central.com on social media. Within a few weeks of launching, Class-Central.com was being used every month by tens of thousands of people around the world.

As more and more universities started offering courses online for free, Class-Central.com grew in popularity. I wanted to work full-time on developing the website, so I applied to a prestigious Silicon Valley EdTech incubator called Imagine K12. To my surprise, I was accepted and they invested $94,000 in Class Central.

Dhawal Shah, founder/CEO of Class Central, which describes MOOCs so that people can make better choices about what MOOCs might be right for them.

The transition was very abrupt. One day Class-Central.com was just a fun thing that I built for myself—the next day it was a start-up with high Silicon Valley expectations. But the only thing I had experience with was writing code. I had no idea how to run a business. I had to quickly learn a number of new skills, including blogging, marketing, managing finances, and project planning, as well as personal development skills like leadership and time management. For some I just winged it or learned on the job; for others I got help from resources like online forums, blog posts, online courses, and MOOCs.

To my surprise, I found out that I was pretty good at some of those new skills. It turns out that with those skills, I was able to reach and be useful to millions of people around the world who are trying to figure out which online course to take. At every stage of the business I have to learn new skills to bring Class-Central.com to the next level. My ability to learn new skills has become my most important skill.

Instructors Are Key

In university classrooms, professors are generally in charge. Sure, certain topics need to be covered, but the instructor decides on both the mechanics and the specifics of delivery—whether to read from notes, perform cartwheels, drone through a PowerPoint, or give exams whenever the moon is full. *No one* second-guesses the professor, especially if that professor is more senior and at an elite university—the very type of professor who is most often asked to make a MOOC.

This traditional "professors make all key decisions" approach has also been taken with MOOCs—everyone in the MOOC production chain defers to the professors' judgment. This can create real problems, basically because most professors are clueless about MOOCs.

There *are* spectacular MOOC instructors who are deeply vested in creating a fantastic learning experience in their online courses. Take, for example, the Ohio State University's Jim Fowler, who makes calculus not only fun but understandable in his almost magically artistic MOOC "Calculus One," and the University of Pennsylvania's Al Filreis, who teaches "ModPo," about supposedly difficult-to-understand modern poetry. These professors have figured out how to take advantage of the medium—Filreis in particular takes a strong interest in connecting with students through live webcasts and active participation on the forums.

But not all professors are that way. For example, after an embarrassing "you can't make this stuff up" implosion in front of 41,000 students, the first MOOC on MOOC-making was halted after a chaotic first week, when students had been given vague instructions for various impossible-to-complete activities.[3] Many other MOOCs aren't bad—they're just bland. The professors simply stand in front of the camera and jaw on and on, rarely bringing in effective visuals or taking advantage of the power of the video medium.

The best online instructors are, of course, experts in their subject area. But they are also receptive to learning at least a little about some of the new technologies that support online learning—screen capture, animation, music, sound effects, video editing, cameras, and more. Part of the challenge in doing a MOOC is simply that they are so new that few instructors have any in-depth experience. As I write this, there are no good books, much less a good MOOC, on MOOC-making. And, of course, most professors have zero training in how to teach, whether in class or online. This means that even diligent, hardworking professors who truly desire to do an excellent job on a MOOC can have difficulty actually doing it.

The take-home message here is that, for a truly outstanding learning experience, you want to look for instructors who have an almost religious zeal for effectively conveying their information in this new online terrain. Again, that's where online reviews are so worthwhile.

Another valuable tool to decide what MOOC to take is this: Research has shown that if you watch a professor on a video for about thirty seconds, you can get a very good sense of how effective that professor actually is as a teacher.[4] Amazingly, even as little as six seconds can allow you to form an effective snap judgment, in part based on emotional micro-expressions too fleeting to truly register. (I sometimes play a quiet guessing game while watching people order coffee about how good they would be as instructors.) One caveat here, though, is that sometimes dryly analytical professors can start out like dreadful bores. But when they begin unleashing their understated but deliciously wicked sense of humor, watch out.

There are people called "instructional designers" who can help professors come up with an engaging MOOC structure. The online "classroom" is different from an in-person class because the tempo is different—good video lessons, for example, are six to ten minutes long. The most helpful instructional designers explain this to instructors and show them how to adjust to this tempo. Poor instructional designers, on the other hand, tend to be more wedded to theory than practice and have little feel for how impractical some theories are. For example, some instructional designers insist that any instructional video should start with bullet points that tell students the key points they will be learning in the video. This approach might have been good in the old days of two-hour classroom lectures, but for a five-minute video, the title alone usually serves as the only bullet point needed. In fact, a list of bullet points at the beginning of a five- or ten-minute video provides a clue that the content will be "pedagogically sound"—and totally snoozeworthy.

Where a professor's internal zeal for creating a superb learning experience really becomes clear is in the quiz questions. It can be tedious and hard to create questions, so some professors farm the work out to teaching assistants. When they do that—though there are some terrific teaching assistants—you're getting the understudy instead of the star. Not cool. And typically not as good a learning experience.

I was told that the best way to do a MOOC was to just "act natural" and speak spontaneously. That just didn't seem right to me—particularly since at the start of the videotaping, my tendency was to freeze and stutter in fear in front of the camera. So instead, I scripted everything and used a teleprompter—no "ums" or "ahs" for me.[5] In the end, students really appreciated the seemingly casual, easy-to-follow performances on the videos. One important point I personally can attest to is that when you're first being videotaped, it can be totally intimidating. No matter how much you try to chase away the feelings, there's always this nervous sense of hundreds of thousands of imagined future viewers. A number of my own first videos (I'm embarrassed to tell you how many) ended up in the computer trash bin.

There's one more reason why the instructors are key: good ones can break through convention and present the material in a fresh, profoundly useful way. Here's what I mean, based on my own experiences with MOOC-making.

A conventional MOOC titled "Learning How to Learn" would have been done by professors from a school of education, not by an engineer and a neuroscientist working on their own. The course would have, in all probability, been geared only for teachers, because teachers of teachers reflexively think that teachers are the ones with a real interest in learning. (For those who think that *any* course on effective learning for general learners would be obvious to make and an instant hit, ask yourself why, out of the thousands of MOOCs available when *Learning How to Learn* came out, no one had ever thought to do a course like it before.)

A conventionally done *Learning How to Learn* course would have been structured with something like two weeks on the history of education, two weeks on theories of learning, two more weeks on how babies learn, with the final weeks touching on how emotions shape our learning and perhaps a bit about deliberate practice and the like. There might even have been a brief lecture or two describing a discipline called "neuroscience." Nothing too deep there, because after all, neuroscience is hard to understand.

Learning How to Learn works because it went back to first principles to present what we know of learning in a fresh, immediately useful way. Neuroscience isn't patched on as an afterthought—instead, it serves as the foundation for the course's key ideas. Where the science gets deep, we use metaphor—we trust in the strength of our learners' abilities to grasp even the toughest ideas when those ideas were presented using the very methods we recommended for learning. We provide direct links to the original research, so learners could check any of our claims for themselves.

How much of what we learn in college could be revitalized by examining the materials in a similarly fresh way? MOOCs offer original, nonconformist professors opportunities for a fresh start, even while providing the platform to reach learners worldwide.

A Sense of Humor

Here's a dirty little secret about learning that you've probably always known, but that nobody really talks about: Even just *thinking* about learning something you aren't partial to, like studying math, activates the insular cortex—a pain center of the brain.[6] Humor can counterbalance this pain—it activates the brain's opioid reward systems.[7] (Yes, humor is like drugging yourself—but in a healthy way.) Humor has an incongruity that makes unexpected neural connections—different types of humor can activate very different parts of the brain.[8] Perhaps, then, humor is the neural equivalent of letting the parts of your brain that are actively involved in learning relax for a moment as other parts of your brain take over to handle the joke. Whether or not this is true, a number of studies have shown that humor is helpful in learning.[9]

Unfortunately, for a lot of people—perhaps especially professors—being funny isn't easy. Coming up with a single humorous bit, especially if it includes animation, can take time and care. Terry and I once got a letter from a fifth grader commending us because she'd never understood that teachers could be so witty. All I could think was, *Of course we were witty. We spent days scripting that "witty," damn it!*

Until recently, when learning was chained to physical classrooms, it was easy to get lazy about being both educational and entertaining: *It's not our job to be funny!* Professors could downplay the importance of humor by claiming there was "too much to cover" to take time for a joke. (Of course, just covering material in class in no way guarantees that students are actually able to learn it.)[10] Or a bit of wit could be dismissed as being a seductive detail that detracts from teaching.[11] But the online world is highly competitive. MOOC-makers who take the time and care to seamlessly integrate humor into what's being taught can not only make learning even tough subjects more enjoyable, but also will make courses that students flock toward.

The bottom line is, when you're looking for a course that will have real value for you, keep an eye out for reviews calling a professor or course "funny" or "witty." It suggests a level of care and creativity, and an understanding of student needs, that may not be in other courses . . . and let's be honest: If there are two courses of equal quality, which would you go for, bland or funny?

Key Mindshift

The Instructor Makes the Difference

Your first impression about whether an instructor is effective is usually spot-on. Look for an instructor who shows unexpected flashes of humor—it's a clue that you'll probably enjoy the time you're putting into learning.

Editing: Every Second Counts

A friend who used to produce TV commercials for a big New York ad agency expressed surprise when she saw some of the footage I'd edited together for my MOOC. At first, I thought it was because she thought I'd done a terrible job. To my astonishment, she complimented me—she said, "People who haven't produced stuff for a living generally go way long. I don't know how you got there, and I mean this in the best possible way—you did this like a TV commercial: making every second count."

How Great Salespeople Are like Great Teachers

Salespeople are also teachers, and time is of the essence. If a prospective client can't quickly grasp what it is you are selling and how it will help them, you don't make the sale and you don't eat. This is especially challenging when you are selling complex technical products and services. We spent hours coming up with metaphors to more quickly and concisely explain our offerings. Imagine if teachers had that kind of moment-to-moment pressure about their students' understanding.

—Brian Brookshire, former director of sales at MortgageLoan Directory and Information, LLC, and also a super-MOOCer

As for how I made every second count, not only did I script succinctly, but I also took care that nothing sat still on-screen for too long.

Even my own talking head was too boring for me to look at for long—which is why I'm first on one side of the screen, then maybe ten seconds later, I'm on the other. Or . . . I'll be standing back far enough that you can see my full standing body, and then I'll suddenly shift to a waist shot, which provides the illusion of looming motion. From an evolutionary perspective, looming motion usually involved critters or objects that could kill you, so we human types tend to snap to attention when something looms, even when it's just on video.[12]

One thing that came as a great surprise was how it would sometimes take me ten hours to edit five minutes of video. (Of course, if I'd been a video-editing professional, I would have been faster.) Although editing was time-consuming, I discovered it is incredibly creative. I began watching television with a different eye—it's fascinating to observe the clever methods used to maintain your interest and to keep a static scene from getting boring.

The best video editors, it seems, have an intuitive feel for the underlying neurocircuitry that keeps people's attention on the video. They do just enough to enhance the message being conveyed, stopping short of distracting from it. Because motion attracts our attention—especially unexpected and looming motion—it is super important in learning online, just as it is in learning in general. This is why some award-winning teachers leap onto desks—and why some of the worst professors inflict death by PowerPoint, with a slew of lifeless images.[13] Unfortunately, despite its importance, video editing is like the ugly stepsister of the MOOC production process—frequently treated as an afterthought.

What good editors do is help design the imagery, sound, and pace of a production. They are co-problem-solvers and understand tricks the professor can use to convey information in the most engaging ways, yet within budget. A good editor is mindful of both the arc of the story or production and of the bits and pieces. The same scene can be dramatically different and dramatically more—or less—engaging depending on how it is shot and cut, and what the audio is. It's exhausting to just watch a tight shot of someone talking, unless perhaps they are as expressive and hilarious as Chris Rock. Few of us are, which is why smart instructors team up with their editors.

Every little bell, whistle, and whizbang that is added to a video to make it more interesting means time and money. If you're a video editor for a MOOC, you typically aren't paid for making an *extraordinary* ten-minute video that might take twenty hours of editing. You're paid for making *a* ten-minute video that might take, at most, a couple hours of editing. And the reality is, many MOOCs just feature the professor seated in front of a bookcase talking, interspersed with a few pictures and some handwriting. Even if you're a fantastic video editor, there's just not that much you can do to add sizzle to a shot of a professor in front of a bookcase.

Closely related to video editing are insights from the world of video gaming. I believe that the best MOOCs of the future will bring much more in from the world of online gaming—not just in the sense of "make it more game-like," but rather, using techniques that video game makers have found grab people's attention and draw them deeply into what's on-screen. Music, sound, motion, humor, game-like design, and human-computer interfaces—all of these play enormously important roles in the learning process that can easily be overlooked because of the trapped-in-a-classroom approaches to learning that we're used to.

The best MOOCs are a meld of academia, Silicon Valley, and Hollywood.

As the price of good video-making equipment keeps falling, it comes within reach for younger and younger students. Some of today's tech-savvy high school kids will go on to become professors who will make fantastic MOOCs—far beyond what we can currently envision. The best MOOCs then, as now, will be a meld of academia, Silicon Valley, and Hollywood.

Crawl Inside a Metaphor

We love metaphors, because they give us a shorthanded way to say something is akin to something else: "Life is a roller coaster" or "Time is a thief." I made liberal use of metaphors in "Learning How to Learn"—even walking around inside them, as with the "pinball in the brain" illustration earlier in this chapter. Unfortunately, many professors are skittish of using metaphor—they perceive it as potentially dumbing down the materials. They're unaware that "neural reuse" theory posits that metaphors use the same neural circuits as the underlying, more difficult concept.[14] The reality is that metaphor doesn't dumb things down for students. Instead, it brings them more rapidly on board with difficult new ideas.

Another reason that metaphor doesn't receive pride of place in teaching is that many MOOC providers and universities rely on learning analytics—statistical data about how people interact with online courses—to drive improvements in their course making. These analytics can point to obvious MOOC-making missteps such as lousy quiz questions, confusingly presented material, and overlong videos that people don't finish viewing. But analytics don't tell you things like "if you'd used a metaphor at the beginning of your theoretical explanation, people would have understood the concept in half the time and had more fun with it, too."

My prediction is that future MOOCs will make much more use of metaphorical visual effects, because MOOCs that use these techniques will tend to be more successful. For you, whatever you're learning, see whether you can make a metaphor to help yourself understand the most difficult topic—you'll be surprised at how much it can bring the key idea to life.

In a related vein, picturing yourself as the same size as whatever you're learning about has a rich history of fueling scientific creativity. Einstein shrank himself five million times to imagine himself traveling so fast that he caught up with a light beam with a wavelength. Cytogeneticist and Nobel Prize–winner Barbara McClintock shrank herself forty million times to enter the realm of the nanometer-sized genes she was studying, where they became like family to her.[15]

In a video, we can actually show the professor riding a light beam. The professor can swim around an alveolar sac, pointing out exactly what happens in your lungs when you are breathing. Instructors can slip onto a proton inside a semiconductor. Sure, you can just show a representation of a light beam, alveolar sac, or proton. But visualizing a person guiding you through the internal intricacies of *anything* can make it more personal and engaging. The "ability to visualize the unseen has always been the key to creative genius," says Einstein biographer Walter Isaacson.[16] Using the power of online video, we can teach ordinary people these exceptional visualizing tools of the imagination.

Formal and Informal Learning Connections

My own sense from interacting with hundreds of thousands of students around the world is that perhaps only 5 to 10 percent of learners are highly self-motivated. These are also the ones who tend to gut it out and do every quiz, project, and other assignment required for formal completion of a MOOC. They also learn on their own quite nicely, thank you very much.

However, perhaps another 60 percent of students learn well *when they're able to also connect with other people to bring the material to life*. Online discussion forums that are built into most online courses make social connections easy. Social media—such as LinkedIn, Facebook, Twitter, and Snapchat—are also used by many MOOCers. Still other MOOCers, as well as libraries, make their own in-person MOOC clubs, kind of like book clubs. Spouses take MOOCs together, and parents enjoy taking MOOCs with their kids. Universities, too, are starting to explore the idea of giving new students a common MOOC experience prior to their orientation. MOOC providers have experimented with meetups and learner hubs, but only an open-source community called Free Code Camp seems to be having any success. (As I write this, Free Code Camp has around a thousand freestanding study groups.)[17] There are also blended learning experiments, where the best learners from a big MOOC with tens of thousands of students are funneled onto an intense, on-campus boot camp–like experience.[18]

On a side note, professors are used to telling students to read textbooks. But even though MOOCs can be as helpful as books for a class, professors just aren't yet used to telling students to go off and watch a MOOC—even though it can mean that the professors have to come to the physical class only half the time for in-person work. (This is the so-called flipped format.) However, I believe that once professors realize that they can teach just as well in half the time by sharing the teaching load with top-notch online materials—well, I think it'll be hard to go back.

In any case, the bottom line is that forming connections with other students is a great way to enhance your learning. Some people like making these kinds of connections, some don't. But even if you're a very independent or solitary type, you might be surprised to find how much you enjoy taking a MOOC with friends or family.

Key Mindshift

Forming Connections

Learning with others can enhance your entire experience—that's why discussion forums and social media can be so worthwhile. Family and friends are often overlooked as learning buddies, but especially when it comes to MOOCs, they can be the most fun!

Where Will MOOCs Take Us?

Over the past century, IQs as a whole have improved dramatically. This rapid IQ ramp-up is dubbed the Flynn effect, after James Flynn, the New Zealand social scientist who discovered it. It's real, not a statistical fluke—most people back in the early 1900s didn't have anything like the learning opportunities we have today to increase our cognitive skills.[19]

Flynn conveys this change with the example of improvements in basketball skill. In the 1950s, when television became a household staple, kids were able to watch top basketball players in action. They saw what the pros were doing, and they brought it to their neighborhood games. As kids started playing against other kids who were just a little better, they got better themselves. And better. And better. There was an ongoing cycle of improvement as kids competed and improved not only themselves but one another. This, in turn, brought higher skill levels to professional basketball.[20]

MOOCs and their ilk are, in some sense, the equivalent of the televised championship basketball games kids began watching. MOOCs can present outstanding teaching by extraordinary teachers, so anybody, student or teacher, around the world, can up their own game. But there's much, *much* more. MOOCs can include little video-editing tricks that grab our attention and unexpected punch lines that leave us laughing—and open to the next tough idea. They can also include metaphors that increase our understanding, as well as splendidly articulated testing systems that reinforce what we need to bone up on and force us to stretch when we reach a plateau.

Essentially, MOOCs are like dating. When you're first going out with someone you're into, you tend to show only your best side. MOOCs, likewise, allow this "best side" selectivity—if the instructor flubs a take, that video can be thrown out and replaced with a better take. Regular classes, on the other hand, are more like a marriage—you see all sides of the instructor. Lousy mood that day? Sorry. Classes are live, and there's no way to take back a poorly taught lesson.

Traditional classes are lacking in comparison to MOOCs in another way. They can be fantastic, but they just don't develop a following that goes beyond the immediate classroom or lecture hall. MOOCs are different. Much in the way of great books—perhaps even more so—

MOOCs can develop a life of their own. The learning that goes on within a MOOC can get caught in an online updraft, spreading to others—locally and around the globe.

The MOOC world is in its infancy. We are at the starting point of a whole new continuous and creative cycle of improvement in teaching and learning, not only for adult learners and university students, but also for secondary and even elementary school levels worldwide. Though, at this writing, "Learning How to Learn" is the world's most popular MOOC, whatever magic we've worked in it will eventually be eclipsed by even better MOOCs—stickier, funnier, and overall "learnier." This will be a great boon for the mindshifts that many people are making to fit into this new era of lifelong learning.

What to Look For in Good Online Learning—and Any Kind of Learning

The easiest way to decide whether an online learning experience is right for you is to check out an online ranking site. Class-Central.com, for example, has a smart way of comparing MOOCs across different platforms by checking out people's reviews. As you try to evaluate the best MOOCs for your purposes, look for the following underlying techniques, approaches, strategies, and tricks, all of which can make a difference in how well you learn, and how much you enjoy the learning process:

- **Metaphor and analogy—wrapped, wherever possible, into visuals and motion.** As "neural reuse theory" shows, use of metaphor and analogy allows you to more rapidly understand difficult concepts.
- **Well-done visuals that relate directly to the material—not generic clip art.** If the instructor can't take the time to develop helpful illustrations, it tells you a lot about how invested that instructor and their in-

stitution are in the course. But just slapping a complex image from the professor's book on the screen also doesn't cut it. People learn differently from videos than they do from books. Complex images need to appear part by part—you'll learn better in the same amount of time if you're not hit visually with everything at once.

- **Lots of motion and rapid cutting.** If intelligently done—and not just style for style's sake—good editing keeps your attentional circuits in gear while *adding* to your comprehension. People are increasingly used to the rapid cutting seen on YouTube videos, where even a simple exhale is sometimes snipped out to create a sense of breathless speed.
- **Humor.** Content that integrates humor and makes you laugh can help activate addictive dopamine circuits of pleasure. It also provides the mental equivalent of a ledge to temporarily rest on and regain your breath when you're climbing an intellectual mountain.
- **Friendly, upbeat instructors.** Look for approachable, encouraging, fun professors who can simplify the material and make the hard stuff look easy. It may seem obvious that all professors should be this way, but they aren't—the reality is, professors *become* professors because they've continually demonstrated how well they can handle hard stuff—or at least stuff that they can complexify and make *look* hard. And okay, truth be told, some professors are pompous blowhards.
- **A minimum of "ums" and "ahs."** Unfortunately, MOOC instructors are typically told to speak "spontaneously" rather than read from a script. Occasional MOOC professors—such as MIT's Eric Lander, in his fabled "Introduction to Biology" course—can pull this off, although even Lander refers to his notes. Many professors stiffen in front of the camera, and their speech suffers accordingly; others simply aren't as polished as they might think they are. You might wonder why all videos in a MOOC can't be "spontaneous," like TED talks. But a typical twenty-minute TED talk demands some seventy hours of

practice.[21] No professor has that kind of preparation time for the hours of footage of a MOOC.

- **Welcoming online environment with mentors and TAs playing an important role.** Mentors and TAs are a bit like park rangers—roaming around online forums making sure the experience is rewarding for everyone, and if necessary, putting out fires. These cheerful assistants also serve as aides-de-camp and lieutenants for the instructors—they often have highly creative ideas about how to improve the course. (This is where having a leader like Princess Allotey, who was willing to solicit and use the best ideas of her team members, is valuable.)
- **Discussion forums and other ways to actively interact with other learners.** Many students benefit from the boost of connections with others. Surprisingly, some of the most introverted learners enjoy discussion forums—it's a way to interact with others even if you're too shy to do much of it in real life.
- **Gamification—bringing in point-scoring, competitive, fun elements to improve learning.** Well-made MOOCs are increasingly taking cues from the gaming world. Games can be addictive—they often are built to give you a series of carefully designed small wins that entice you to go deeper and deeper into the material. (*Wait . . . lunch was two hours ago?*) Zippy music and sound effects at the right moments can enhance feelings of immersion.
- **A well-designed, easy-to-follow course structure.** A glance at the syllabus and the layout of the course on the webpage gives a sense of whether the course is for you. If reading descriptions of the course materials makes you curious to learn more, it's a good sign.
- **Quizzes.** One of the best methods for making sure you really know the material is to test yourself every chance you get. Online quizzes make this process easier. Plus, well-made tests strengthen your retention of the most important aspects of the material. Be careful if reviews of a course mention problems with the quizzes.

- **A final project.** It's uncanny how we can remember that we did a project or report for a class, years after we've forgotten virtually everything else. Not only that—a good final project can leave you with real love for the material. (I once met a man who moved to Pennsylvania because he'd fallen in love with the state after doing a report on it in grade school.)

 **Now You Try!**

Find a MOOC!

Go online and find a MOOC on a topic you're interested in. The easiest way to do this is to go to Class-Central.com to run a search. Class-Central allows you to build your own list of courses and follow universities or topics, and to receive e-mail notifications of upcoming courses and most popular courses.

When you are searching MOOCs of interest, you'll need to be a little careful. The subject matter in MOOCs is so broad that you may not even think to look for a MOOC on your favorite minor novelist or television drama—though such a MOOC may very well exist.

It's helpful to know who some of the major MOOC and online learning "players" are. This listing is of U.S-based and university-affiliated providers unless otherwise noted. The word "MOOC" is used loosely to mean any low cost or free online course.

- **Coursera:** The largest MOOC provider. Has courses on many different subjects and in in many different languages. Also offers an MBA and data science master's degree, and offers "specializations"—clusters of MOOCs.

- **edX:** Has a large number of courses on many different subjects and in in many different languages. Offers "MicroMasters"—clusters of MOOCs.
- **FutureLearn:** Has a large number of courses on many different subjects and languages, particularly, but not exclusively, from British universities. Offers "Programs"—clusters of MOOCs.
- **Khan Academy:** Offers tutorial videos on a large number of subjects, from history to statistics. The site is multilingual and uses gamification.
- **Kadenze:** Special focus on art and creative technology
- **Open2Study:** Australia-based, many subjects
- **OpenLearning:** Australia-based, many subjects
- **Canvas Network:** Designed to give professors an opportunity to give their online classes a wider audience. Has a large number of courses on many different subjects.
- **Open Education by Blackboard:** Similar to Canvas Network.
- **World Science U:** A platform designed to use great visuals to communicate ideas in science.
- **Instructables:** Provides user-created and -uploaded do-it-yourself projects which are rated by other users.

Here is a listing of professional- and specialist-oriented platforms (some are subscription-based):

- **MasterClass:** Top masters teaching their subject—Kevin Spacey teaches acting, Serena Williams teaches tennis, Christina Aguilera teaches singing, Annie Leibovitz teaches photography, etc.
- **Udacity:** Tech-related courses for professionals, offers "Nanodegree programs," and a master's degree in computer science through Georgia Tech.
- **Lynda.com/LinkedIn Learning**: Thousands of courses in software, creative, and business skills.
- **Codeacademy:** Free coding classes in popular computer languages.

- **Shaw Academy:** Ireland-based. Courses on many professionally related subjects are conducted live at convenient times—students can interact with the instructors as well as with fellow students.
- **Pluralsight:** Online developer, IT, and creative training—large library of courses. (Royalties created the first millionaire teacher from online teaching.)
- **Udemy:** Courses by self-claimed experts on a wide variety of topics, including technical topics and job-related skills. Popular with corporate trainers.
- **Stone River Academy:** Web, app, and game development
- **Skillshare:** Courses by self-claimed experts in creative arts, design, entrepreneurship, lifestyle, and technology.
- **Eliademy:** Finland-based, simple platform for anyone, for example, K–12 teachers, to create, share, and teach online courses.
- **Treehouse:** Courses on web design, coding, business, and related subjects.
- **General Assembly:** Courses on design, marketing, technology, and data.
- **Tuts+:** How to tutorials

There are also MOOC and online learning platforms that specialize in certain languages and cultural spheres (there is some overlap between areas and languages). These are some examples:

- **Arabic-speaking world:** Rwaq, Edraak
- **Austria:** iMooX
- **Brazil:** Veduca
- **China (simplified Chinese):** XuetangX, CNMOOC, Zhihuishu
- **Europe:** EMMA (European Multiple MOOC Aggregator), Frederica.EU
- **France:** The France Université Numérique, OpenClassRooms, Coorpacademy

- **Germany:** openHPI, Lecturio, Moocit, Mooin, OpenCourseWorld
- **Greece:** Opencourses.gr
- **India:** SWAYAM; NPTEL
- **Italy:** EduOpen, Oilproject
- **Japan:** JMOOC
- **Russia:** Stepik, Intuit, Lektorium, Universarium, Openedu.ru, Lingualeo .com
- **Spanish- and Portuguese-speaking world:** Miríada X, Openkardex, Platzi
- **Sri Lanka:** Edulanka
- **Taiwan (traditional Chinese):** eWant
- **Ukraine:** Prometheus

Also worth noting:

- **Duolingo:** free language learning app for many languages.
- **Crashcourse:** A series of witty educational videos that have expanded from the initial humanities and science (YouTube).
- **VSauce:** Incredibly funny and quirky educational videos (YouTube).

Jot your MOOC ideas down under the title "Possibilities for Expanding My Learning."

Chapter 13

Mindshift and More

"Louise" had a problem: Her horse, "Specs," was trying to kill her.[1]

He had just kicked her in the head, knocking her flat. It took five minutes before she came around—fortunately, after she'd hit the dirt and stopped moving, Specs had lost interest and wandered off.

Louise had spotted a flyer for Specs on a gas station bulletin board when she and her husband had driven from their coastal Washington State home to eastern Washington to visit relatives. Specs's owner, a rancher, had described him as an adorably curious little guy who loved investigating anything new, even if it meant stepping hoof by hoof into a new water trough or managing to get himself entangled inside a brand-new tent. There was just something about the description that captured Louise's heart. She felt certain Specs would become her perfect horse—the quiet hobby she needed for retirement after her many years of work as a part-time secretary and mostly stay-at-home mom. On their way back to western Washington, Louise and her husband arranged to meet Specs.

Strangely enough, Specs pretty much ignored her. When she'd led him out of the corral, he'd charged into the lead, dragging her along as he snatched clumps of grass from the edges of the path. Louise already had him in her heart, though—they bought him on the spot. Sure, Specs was a little crude. But with a little training, Louise was sure Specs would be Mr. Ed in no time.

It wasn't turning out that way. One afternoon, Specs had reared overhead to strike—Louise could see all the way to his back teeth as he leaned in to bite. Yet another time, Specs kicked Louise all the way out of the barn. She'd hit a plank and spent the next few weeks limping. Her list of injuries went on and on—thumb cut to the bone, bruises, stomped toes.

Louise tried to ride Specs, but he would buck her off as soon as she mounted—or he would wait until she was more relaxed and then flop on his side and try to roll over on her. If she led Specs on a halter, he would wait until they came to a steep hillside and then head butt her down the slope. Or he would just take off and gallop through the neighbors' yards.

Louise had loved animals all her life—she'd always been fascinated by how they think and learn. But this situation was becoming increasingly out of control. In fact, Louise was beginning to suspect that Specs was something of an equine psychopath.

But there was a problem. If Louise told anybody about what was really going on, Specs could end up dog food.

She felt trapped. And Specs was getting worse.

Discovering Hidden Potential

Modern humans are thought to have headed into Europe and Asia some sixty thousand years ago, where they found modern horses waiting for them.[2] Dinner! Humans hunted, killed, and ate equines with abandon for tens of thousands of years. Finally, about six thousand years ago, people began realizing there was hidden potential in horses.[3] They could be milked. They could carry and drag stuff. They could even—wow!—be ridden. The domestication of the horse had a profound effect on the course of civilization. Earlier in *Mindshift*, we saw echoes of the power of horses in the extraordinary spread of the Comanches.

Consider what this means: It took humans more than fifty thousand years to discover the extraordinary hidden potential of the horses *hidden in plain sight right in front of them.*

This book's subtitle is *Break Through Obstacles to Learning and Discover Your Hidden Potential.* This covers enormous scope, as we've seen while meeting people around the world, from many walks of life, who've made mindshifts. But, especially when we delve into the science, a common thread becomes clear: People can often do more, change more, and learn more—often far more—than they've ever dreamed possible. Our potential is hidden in plain sight all around us.

I was inspired to write this book because, like Dutch online gamer Tanja de Bie, I am a midlife second chancer. Long after I was supposed to be set in my career and set in my ways, I was lucky enough to have the opportunity to transform myself from a person whose only apparent talent was in language and the humanities. This allowed me to turn to a new career trajectory that eventually led to my becoming a professor of engineering.

These days, working behind the scenes running the MOOC "Learning How to Learn," I'm newly inspired by all the changes that learners are making. Over and over, I see that people have the ability to learn and change at every age and stage—not only from the humanities to engineering, as I did—but in virtually any direction. This sort of mindshift isn't just about following your passions, but about *broadening* your passions—reenvisioning yourself in new directions, both in your private life and in your career, and then taking the steps as a learner to expand your horizons.

I've heard thousands of inspirational stories while writing this book. What this should tell you is that the biographical vignettes I highlight in *Mindshift* give just the tiniest glimpse of what's possible. *Mindshift* could easily have had ten times the stories I included, but it would still be tied together by the common thread of people using learning to re-shape their work and their lives.

There are two overlapping aspects of learning, and we've explored both in this book. The first is to realize that mindshift—deep changes in life that occur through learning—is something that can be done at any age, with any goal in mind. Books and other ways of learning can help spur a series of changes—as we saw with Claudia Meadows's escape from depression and Adam Khoo's success in professional and personal life through retooling of his attitude. Super-MOOCers show us how powerful, fun, and even addictive learning can be. It also keeps our minds fresh as we age. Indeed, the retirees I've met who have a zest for learning remind me of extraordinarily mature intellectual teenagers—the kind of people who are just plain fun to hang out with.

The second aspect of mindshift relates to career: career selection, career enhancement, and career switching. Each of these demands not only a hunger to learn, but also an ability to look dispassionately at the direction and goals of your learning. Sometimes, as with physics-major-turned-neuroscientist Terrence Sejnowski, it can be worthwhile to step back and assess the big picture of a discipline. Upon recognizing the limitations of his particular branch of physics, he turned to neuroscience, where he was able to make far more significant contributions. In contrast, master marketer Ali Naqvi landed in a field—search engine optimization—where his limited computer-related skillset was a problem. He stepped up, using MOOCs to fill the gaps in his expertise. His growing competence resulted in promotion after promotion, which allowed him to move rapidly into management.

It's important to remember that both Terry and Ali found that their seemingly irrelevant past provided value in their new careers. Terry's training in physics gave him the underpinning for the mathematical models he uses in neuroscience. Ali's background in golf gave him the emotional awareness to avoid letting past mistakes affect his future actions, as well as a special aptitude for sports-related marketing.

In fact, a common theme throughout this book is that background and training from the past that at first seem entirely useless often prove valuable in your new job. For example, Arnim Rodeck's analytical way of thinking, which grew from his training as an electrical engineer, enhanced his later-life shift to woodworking. Tanja de Bie's seemingly frivolous experience in gaming resulted in an awesome job managing online communities. Jonathan Kroll's background in Romance languages helped him learn computer science. And musician-turned-med-student Graham Keir was able to make more effective medical diagnoses because of his musical expertise.

Graham Keir's unlikely shift from his beloved music to the hard science of medicine also illustrates how we can broaden seemingly invincible "this is the only life for me" passions to encompass entirely new passions—even in areas we had previously despised. The increasing availability of online learning tools makes this more doable than ever. Graham, as we saw, started his mindshift by looking at a simple precalculus e-book on his iPhone, which allowed him to run through concepts while riding the bus to performances or to school. Others are using digital tools and MOOCs to enhance their knowledge in their specialties, or to second-skill—to explore new career possibilities and general interests.

What's so beautiful about the online learning world is that it *lends itself well to how the brain learns.* With MOOCs, for example, teaching can be condensed into bite-size, highly memorable videos that grab your attention. Every video lecture can be the *best* lecture a professor has ever given. Powerful online tools can allow you to practice over and over again until each concept is chunked and becomes second nature. When coupled with conventional textbooks and even "in-person" guidance with activities within a classroom, online learning can become part of the "best of all worlds" of learning.

The socially connected learning communities of MOOCs provide another bonus. As MOOCs mature, these communities will continue to improve. And MOOCs are indeed maturing—in the previous chapter you got enough of a peek behind the scenes of MOOC-making to perhaps sense the direction of the best MOOCs future.

What's particularly encouraging is that thoughtful, creative, broadly available online teaching not only improves the lives of students, but also calls for teachers to up their game. The new teaching materials are also kicking off a digital learning revolution in the ed-tech scene, which is in turn reinvigorating the education-learning sector. Of course, there's even more going on with the trends in do-it-yourself tinker-type learning, most prominently including "makerspaces" in libraries and community centers where people can access 3-D printers and various tools.

For every job . . . the number one thing we look for is general cognitive ability, and it's not IQ. It's learning ability. It's the ability to process on the fly. It's the ability to pull together disparate bits of information.[4]

—Laszlo Bock,
senior vice president of People Operations at Google Inc.

Part of the challenge with mindshift is that, early on, *most of us aren't taught how to learn.* This means that in our youth, we frequently fall into some pursuit that, at least at the time, we feel we're better at. Then we presume that that's our passion—it's what we should do. This thinking is reinforced since our grades tend to suffer when we wander outside what we're "naturally" good at. We tend to forget that some things take longer to get good at—and as we do get good, those things can, in turn, become new passions. There's more: As mathematics educator Princess Allotey has shown, when our passions are temporarily blocked by unlucky twists of fate, we can use that time, not just to develop broader passions, but also to become more well-rounded human beings. Princess's public speaking abilities, as well as her ability to move past her feelings of imposterhood, will serve her in good stead throughout her life.

Since access to standardized learning was almost entirely limited to brick-and-mortar schools designed for the young over the course of the past several centuries, societies fell into a sort of "learning is only for the young" mentality. But thanks to MOOCs and other online learning opportunities, people are beginning to recognize that learning is for everyone, at all stages of life. This is why innovative countries like Singapore are putting emphasis on learning lifestyles that value learning of any kind, no matter what the subject or goal.

Knowledge of the brain's workings can allow us to leverage every aspect of our learning. In this book, I've tried to convey some of the latest hopeful insights about how adults can keep learning and growing well into advanced maturity—and how a learning lifestyle helps prevent the mental stagnation and decline we often associate with old age. The digital medium is part of this. For example, as researchers Daphne Bavelier and Adam Gazzaley have shown, video games can provide fantastic new ways of not only maintaining but enhancing our cognitive abilities. But nondigital approaches such as meditation may also enhance various aspects of the learning process. We've seen how focused-attention forms of meditation boost the neural networks involved in our ability to concentrate, while open monitoring may improve the diffuse, imaginative processes related to the default mode network.

 **Now You Try!**

Key Ideas in *Mindshift*

Under the title "Key Mindshift Ideas," make a list of what you thought were the key points of this book. (This will help you to chunk and remember those ideas.) Do you think others' lists would look like yours? Why might other lists differ?

Owning the Situation

Other mammals seem to learn in many of the same ways we humans do—they even show evidence of using focused and diffuse modes.[5] It's just that a typical mammal's inability to use speech can make learning a lot tougher. Think of a dog as it bounces around trying to guess what you want—*Did you want me to roll over? No . . . How about to sit up? Drat . . . that's not it either. Please—just tell me what you want me to do and I'll do it!*

Not being able to communicate, it seems, was part of Specs's problem.

Specs was a jet black horse with a sprinkle of white spots on his rump—hence the name Specs, short for Speckles. He'd had a tough birth and had been sickly his first month of life, but he was an adorable little scrapper who quickly became something of a pet. The rancher's teen daughter, Edwina, had planned to keep him, so she began to teach him tricks she'd learned from an old ranch hand. One of the first tricks Edwina taught him was to lie down. Unfortunately, the way she did it was by repeatedly kicking Specs's left leg and yanking his head around, which made him lose his balance and fall over.

Louise, reflecting on this sort of training, said, "Some people think teaching tricks to a horse is cute, but it has to be done correctly because what you teach a horse, they *own*." In other words, seemingly minor tricks can become a fundamental part of a horse's way of interacting with humans.

And Specs *did* own this particular trick. Lying down became his default behavior. Whenever he got stressed, he'd just drop to the ground—after all, that's what people had apparently wanted him to do. In any case, Specs had discovered that lying down, no matter what the provocation, usually made any unpleasantness stop. Even better, it gave him power to control the people around him. For example, if someone was riding Specs and he didn't feel like being ridden, all he had to do was stop, drop, and roll. That put an end to being ridden, straightaway.

Lying down wasn't the only thing Specs learned from Edwina. Edwina used tools to elicit the behavior she wanted—pain-causing tools. To get Specs to back up, she would thump his chest with a metal hoof pick. He would back up, but it was clear what he'd learned: *I will back up now because you are annoying me, but don't expect me to do it if you don't have that stupid hoof pick hitting me!*

Among the behaviors that horses display when they are frustrated and don't respect or trust whom they're dealing with are kicking, biting, and stomping. Specs used them all. Edwina had other horses that were bigger and more tractable than Specs, so she ended up riding them and leaving Specs to his own devices. Specs wasn't big enough for regular farm work, so Edwina's father packed Specs off to a riding program for handicapped children. Horses for these children need to be very gentle and show a great deal of patience. That wasn't Specs—so he ended up back at the ranch.

Horses can learn behavior, like lying down, that can be either good or bad. Edwina hadn't meant to be hurtful in her attempts to train Specs—she'd just followed the ranch hand's recommendations. But due to the way Edwina had treated him early on, Specs exhibited an attitude of resentment toward learning and toward people themselves. He didn't understand what he was being taught. If he could have talked, it almost seemed he would say, "The way I'm treated is so unfair!" It was clear that, from his perspective, learning was annoying—in fact, *people* were annoying.

This feeling that learning is annoying, to be done only in response to being prodded, isn't seen just in horses—it's in people, too. Early on, high school dropout Zach Caceres watched as many of his friends, turned off from learning, fell into far more problematic behaviors. Of course, Zach's friends weren't locked in a stall or tied to a post—they had more options than Specs. So instead, they messed around in class, tuning out teachers they had no respect for, only doing the bare minimum to pass (echoing Specs—*I will do it now, but don't expect me to do it if I don't have to!*). Zach's friends soon got into drugs or found that violence worked to get them what they wanted. It was the human equivalent of Specs's road to hell.

The Breakthrough

Back when Louise was a child in rural Forks, Washington, fifty years ago, she'd had a sweet and gentle horse that had happily ferried her around. This half-century-old experience was part of the problem—Louise thought she knew horses. The truth was, with a cantankerous critter like Specs, she was a sixty-two-year-old neophyte, in way over her head. Nothing she tried was working. Specs's belligerent behavior—pushing back, kicking, biting—could have been handled by an experienced horsewoman who would have whipped him into shape in no time. But for Louise, Specs's aggressive behaviors were just scary and upsetting. In any case, whipping a horse into shape wasn't Louise's style.

In desperation, Louise set her horse-training books aside and reached out online to a horse-training expert. (In other words, like many we've seen in *Mindshift*, she found herself a mentor.) Louise says: "The trainer would send me assignments and I would take video of ourselves doing them. Man, she was tough—I got yelled at many times for my mistakes, all because she was concerned about my safety. She worked with me for two years. I kept quiet about it because no one would have believed that I was trying to solve my problems by working with someone who lived on the opposite coast—but we did it! I owe her a great deal—thank goodness for online learning."

Louise was startled to discover, first off, that Specs hadn't learned to respect a person's personal space. For example, when Louise would go into the corral to spend time with Specs, he would push right up to her chair, nosing her glasses askew, grabbing her book with his teeth, even knocking over her chair as she jumped aside to avoid him. He had never learned patience, so he would push her aside to get what he wanted—usually food. Specs had also learned that if he bit, reared, balked, or lay down, he could get his way. Of course, all this made him very unsafe for Louise to be around.

Louise wondered how she could even begin to tackle all these issues. But she eventually had a breakthrough. It came when the horse trainer taught Louise the intricacies of the simple "bridge and target" way of communicating with horses.[6] In brief, "bridge and target" involves teaching an animal the *target* that you want the animal to move toward. (In Specs's case, it was a two-foot plastic disk with an *X* on it.) The *bridge* to get the animal to the target is a sort of clucking sound you make as the animal approaches the thing—the nearer the animal gets, the faster you cluck. Bridging is similar to the hot and cold game that children play, using a faster clucking sound that communicates *warmer, warmer, that's it!*

Just as teacher Anne Sullivan found a way to reach blind, deaf, and unruly Helen Keller by signing words on her hand, bridge and target finally provided a way to get through to Specs. Through this technique, Specs discovered that if *he himself* decided to head toward a target (active learning!), he not only could get a sign from a person about whether he was on the right track, but he could get a blueberry treat when he got there. *He* was the decider! There was nobody pushing him around with metal hoof picks or making him fall over.

The trainer taught Louise to start *watching* Specs—really watching, so she could read his attitude and demeanor. By observing him, Louise saw that though Specs would do what she requested, he would sometimes do it with his eyes narrowed, his ears pressed back angrily, and his body tensed in a sort of *I'll do it, but screw you* way.

Louise observes that the trick to reading an animal is to look for the thought behind the behavior, not just whether they're accomplishing the task at hand. For example, there's the kid who finally cleans his room after he's told to do it umpteen times. Sure, he does tidy up—but only after stomping off with a disrespectful "FINE!" and then mumbling bad things about his mom under his breath.

Well, doing the job with a bad attitude isn't quite the same as doing the job. Louise notes that "Attitude trumps everything. You have to be very conscious of what you are rewarding."

Key Mindshift
Attitude

Attitude trumps everything.

Reading attitude starts with remembering to look for it, Louise advises. "You actually can feel when a horse is happy and relaxed, as opposed to when they are tense and something is wrong."

This is the part that is hard to get from a book, Louise observes. "It is the intuitive part of learning that comes with experience *and* a good mentor to point these things out."

Another invaluable aspect of Louise's learning was videotaping and evaluating her performance. As Louise explains, "I would think I was hot stuff as a trainer," but then her mentor would look at the video and tell her what was really going on.

When Louise first started training Specs, at the trainer's direction, she kept herself behind the fence for a few weeks. Specs's first lesson was simply to learn to keep his head on his side of the fence—while maintaining a happy attitude: his ears forward (generally a positive sign in horses) and his body relaxed. Doing that would get him a reward. Louise did punish Specs for rudeness—but his punishment was simply that she left, which, in turn, meant that he lost the opportunity to get treats. Adding a cost for bad behavior increased Louise's significance in Specs's eyes, because if he didn't play by the rules, she left—game over. Oddly enough, this also meant that Specs had control over the situation.

Louise gradually began working with Specs in the corral. She learned to watch for and encourage exhaling, which, just like in humans, is a sign of emotional and physical relaxation. She says, "If either one of us is anxious, I will audibly exhale a big calm breath, and he will often do the same." She explains that doing this—simply breathing—changes the emotional picture for both of them.

As Louise and Specs's mutual language developed, Specs began to learn more and more. He learned how to play soccer, fetch, paint pictures with a brush, zoom past Louise in a round of "keep away from the barn," and play the piano Liberace-style by sliding his nose zestfully over the keyboard. Louise would watch from the window of her kitchen and see Specs practicing lessons on his own, and even creating new ideas to show her.

Specs now *loves* having his hooves trimmed, reaching out to nibble Louise affectionately as he extends a hoof out as if he's getting a manicure at a spa. Louise can now ride Specs without a saddle or bridle—she simply lets him know by her voice which way she would like them both to go.

So, ultimately, bridge and target provided a language that Louise and Specs could use to communicate. Even better, the communication allowed Specs to save face—yes, it seems even horses have a sense of personal pride—and to enjoy success. Specs now had the ability to control his environment in positive ways and get rewarded for it.

Looking back on the progress the two of them made, Louise observes, "If there's trust and respect with any animal and you find a way to communicate with them, they will communicate back. And you will begin to find hidden levels of potential there."

Louise herself, like Specs, has undergone a mindshift. With two sisters who are teachers, she can't help but reflect on the human parallels to Specs's transformation—that teachers in public schools often don't have students' respect or trust, and that there is limited punishment allowed as consequences for impolite, disrespectful, or dangerous behavior.

Louise says, "Specs rebelled against what he did not understand or thought was unfair. Most horses just accept and don't challenge. If Specs is different from many horses, part of that difference is that trying to pressure him just created more fight in him. Sometimes, even today,

Specs will test the limits with an old behavior such as nipping. However, he accepts the correction calmly, like a child busted when he tests his parents."

Today, to see Specs and Louise together is to marvel: The pair have a clear respect and love for each other. Specs isn't just interested in the treats and attention that Louise offers. *He loves the ability to learn.*

Louise recently added the idea of chunking to her teaching—she finds that after three repetitions, Specs usually gets the idea of some new task or trick, whether it's to pick up and paint with a paintbrush, hop up on a pedestal, or close a gate. She marvels as Specs uses deliberate practice to perfect his skills—perhaps in kicking a ball toward a net, picking up a rubber baton to drop it through a hoop, or cantering in a tight circle (not necessarily easy for a horse). Louise feels that Specs truly wants to master what he is learning. What's particularly fun for her is how human-like Specs is in his learning: He'll struggle with something new, but, each time, he comes back to it better than when he'd left off.

To Louise's amazement, Specs isn't just a student—he's an innovative creator who enjoys coming up with new ideas. As Louise has discovered, "When you develop a sophisticated learner, as Specs is, he will show his creativity to please himself, not you. Part of Specs's creativity is that he has learned how to manipulate me to get what *he* wants. This is very different than his just learning tasks that I dream up." So, for example, when he is ready for company, he will stand on the highest pedestal in the corral and give a single whinny of a certain tone, meaning, "Come out for a visit; maybe some hay please." In other words, Specs has managed to train *Louise* to come in response to his call. And Louise listens and responds to Specs—like when he shows opinions about some of the lessons, sometimes offering his own variations. Oddly enough, like a great teacher, Specs has a mirthful side to him—he really loves making Louise laugh. Her joy seems to give him great satisfaction and pride, empowering him as both teacher and learner. His joy affects Louise similarly.

Louise looks back on her path—from being in tears every night to coming to see Specs as a very special gift in her life. She says, "I believe that when Specs began to understand his world, and he discovered he could actively manipulate it in healthy ways, that changed his attitude." She's excited about how far she and Specs will go and, she says, "how much more there is to learn" for both of them.

Far from being an equine psychopath, Specs is something of a horse genius. Whether it's trying to climb into a car with his people, or letting himself in and out of the corral at night to sneak into the house, he's always game to explore new worlds—and to show off what he knows for anyone there to watch.

Opening the Door

If you've read this book, you undoubtedly have a thirst for knowledge. I'm hoping what you've read has widened the scope of what you thought you could do, helping you broaden your passion for discovery. Remember that it took nearly fifty thousand years for humans to see the use of horses under our very noses. How many insights today are right in front of you that could make a dramatic difference in your life once you discover them? Learning can be an overwhelming pursuit, but it also provides a way for us to meet some of our deepest needs to live as full, vibrant beings.

However, like Specs, many rebel against learning, or resign themselves to staying where they are in life. What about people like these, you might ask? How can they make a mindshift?

If I can leave you with one last message in this book, it is this: Sometimes it just takes that one special person, a mentor, to unlock—or reframe—the doors, as Louise has done with Specs. I hope this book has inspired you to turn to others—to those who are shut down. May your own discoveries open the minds of those you touch, so they, too, can discover the beauty and joy of learning.

 **Now You Try!**

Mastering Your *Mindshift*

Now is the time to review your notes and thoughts in relation to this book. We've covered many areas, but your observations should fall into distinct categories: broadening your passion, creating your dream, mind tricks for success, and of course, many others. When you read through your notes and reflect on your thoughts, what common threads about yourself, your goals, and your dreams do you perceive in your writing? Under the title "Mastering My Mindshift," write your synthesizing thoughts about your personal breakthroughs and epiphanies. What are your concrete plans now as a result of what you've discovered about yourself through the course of reading this book?

One final question. Through your reflections, you have undoubtedly found a positive path toward your future. Is there a way to start someone else on a positive path as well?

Acknowledgments

It's hard to know where to even begin in thanking all the great people who have helped this book come to life. Special appreciation to Joanna Ng, my editor at TarcherPerigee/Penguin Random House, whose incisive edits and big-picture direction have made a powerful impact on how this project has unfolded. My thank-you as well to Sara Carder, editorial director at TarcherPerigee/Penguin Random House, whose behind-the-scenes guidance and input has been invaluable. No author could be luckier than to have a literary agent of the caliber of Rita Rosenkranz. Being able to team with Rita is one of the luckiest perks I have had as an author.

Amy Alkon is a fantastic science writer, editor, and terrific friend who has combed through every word of the initial draft of this book, making it far better in the process. I'm deeply grateful she was willing to share her talent even while working on her own upcoming book. Friends do not get any better than Amy Alkon.

Great appreciation to Cristian Artoni, Daphne Bavelier, Pat Bowden, Brian Brookshire, Zachary Caceres, Jason Cherry, Tanja de Bie, Ronny De Winter, Adam Gazzaley, Alan Gelperin, Soon Joo Gog,

Charles G. Gross, Paul Hundal, Graham Keir, Adam Khoo, Jonathan Kroll, "Hans Lefebvre," "Louise," Claudia Meadows, Ali Naqvi, Mary O'Dea, Laurie Pickard, Arnim Rodeck, Patrick Tay, Ana Belén Sánchez Prieto, Geoff Sayre-McCord, and Terrence Sejnowski, whose insightful e-mails, essays, and personal discussions helped form the basis for their chapters or sections and whose comments frequently improved the book as a whole.

Special thanks also to Charlie Chung, Sanou Do Edmond, Stephanie Caceres, Wayne Chan, Jerónimo Castro, Yoni Dayan, Giovanni Dienstmann, Desmond Eng, Beatrice Golomb, Jeridyn Lim, Edward Lin, Vernie Loew, Chee Joo "CJ" Hong, Anuar Andres Lequerica, Hilary Melander, Mary O'Dea, Patrick Peterson, Emiliana Simon-Thomas, Alex Sarlin, Mark Smallwood, Kashyap Tumkur, Brenda Stoelb, David Venturi, and Beste Yuksel.

And most of all, I want to thank my wonderful family. My son-in-law, Kevin Mendez, has always been there whenever I needed artistic insight. He's also been a fountain of wisdom with relation to relevant reading materials. My Kosovar son, Bafti Baftiu, and my granddaughter, Iliriana, have given great hugs and encouragement. My daughter Rosie Oakley is as good an editor as she is a doctor—which means I'm very lucky to have had her help. My daughter Rachel Oakley is always there for me, both with inspiration and photographic acumen. My brother Rodney Grim is a family mainstay.

And finally, I can't help but think I'm the luckiest woman alive to have met and then said yes to Philip Oakley when he asked that we join our lives together. He is the beacon for my soul and lodestar for my spirit. This book is dedicated to him.

Illustration and Photo Credits

1-1 Photo of Graham Keir courtesy Graham Keir.
1-2 Photo of a Pomodoro timer by Francesco Cirillo rilasciata a Erato nelle sottostanti licenze seguirÃ OTRS, available at http://en.wikipedia.org/wiki/File:Il_pomodoro.jpg.
2-1 Map of Seattle, Washington, USA, derived from the map available at https://commons.wikimedia.org/wiki/File:Blankmap-ao-090W-americas.png.
2-2 Photo of Claudia Meadows © 2016 Susie Parrent Photography.
3-1 Map of Ali Naqvi's travels derived from the world map available at https://commons.wikimedia.org/wiki/File:BlankMap-World-v2.png.
3-2 Photo of Ali Naqvi courtesy Ali Naqvi.
3-3 Light microscopy image of neuron with new synapses © 2017 Guang Yang.
4-1 Map of the Netherlands derived from the world map available at https://commons.wikimedia.org/wiki/File:Netherlands_(orthographic_projection).svg.
4-2 Photo of Tanja de Bie courtesy Barbara Oakley.
4-3 Boys and girls have similar math abilities © 2017 Barbara Oakley.
4-4 Boys and girls have different verbal abilities © 2017 Barbara Oakley.
4-5 Boys and girls have similar math abilities and different verbal abilities © 2017 Barbara Oakley.
4-6 Photo of Kim Lachut © 2016 Kim Lachut.
5-1 Map of Zach Caceres's travels derived from the world map available at https://commons.wikimedia.org/wiki/File:BlankMap-World-v2.png.
5-2 Photo of Zachary Caceres © 2017 Philip Oakley.
5-3 Photo of Joan McCord courtesy Geoff Sayre-McCord.
6-1 Map of Singapore derived from the world map available at https://commons.wikimedia.org/wiki/File:Blankmap-ao-270W-asia.png.

6-2 Photo of Patrick Tay courtesy Patrick Tay.
6-3 Illustration of Singapore's tripartite approach © 2017 Barbara Oakley.
6-4 "T" image © 2017 Kevin Mendez.
6-5 "π" image © 2017 Kevin Mendez.
6-6 Second-skilling © 2017 Barbara Oakley.
6-7 Mushroom © 2017 Kevin Mendez.
6-8 Talent stack of skills © 2017 Kevin Mendez.
6-9 Dr. Soon Joo Gog © 2017 Barbara Oakley.
7-1 Photo of Adam Khoo courtesy Adam Khoo.
7-2 Mind map of weathering courtesy Adam Khoo.
7-3 Balloons of thinking modes © 2017 Jessica Ugolini.
8-1 Photo of Terrence Sejnowski as a young man courtesy Terrence Sejnowski.
8-2 Map of Terrence Sejnowski's travels for study and work derived from the map available at https://en.wikipedia.org/wiki/File:BlankMap-USA-states.png.
8-3a Midline frontal theta waves © 2017 Kevin Mendez.
8-3b Front to back theta waves © 2017 Kevin Mendez.
8-4 Photo of Terrence Sejnowski at Waterton courtesy Terrence Sejnowski.
8-5 Photo of Alan Gelperin courtesy Alan Gelperin.
9-1 Map of Accra, Ghana, derived from the map available at https://commons.wikimedia.org/wiki/File:Ghana_(orthographic_projection).svg.
9-2 Photo of Princess Allotey courtesy Princess Allotey.
10-1 Map of Arnim Rodeck's travels derived from the world map available at https://commons.wikimedia.org/wiki/File:BlankMap-World-v2.png.
10-2 Photo of Arnim Rodeck © 2016 Arnim Rodeck.
10-3 Photo of Arnim's woodwork (Galiano Conservancy Association) © 2016 Arnim Rodeck.
10-4 Photo of Arnim's woodwork (front doors) © 2016 Arnim Rodeck.
11-1 Photo of Arnim's notes and books © 2016 Arnim Rodeck.
11-2 Photo of Laurie Pickard © 2015 Laurie Pickard.
11-3 Photo of MOOC certificates © 2017 Jonathan Kroll.
11-4 Photo of Ana Belén Sánchez Prieto in the studio © 2016 Ana Belén Sánchez Prieto.
12-1a Photo of Barb in the basement © 2017 Barbara Oakley.
12-1b Composite photo of Barb in the basement © 2017 Barbara Oakley.
12-2 Photo of Terrence Sejnowski running on the beach © 2017 Philip Oakley.
12-3 Photo of Philip Oakley in the studio © 2017 Rachel Oakley.
12-4 Photo of Dhawal Shah © 2016 Dhawal Shah.

References

Ackerman, PL, et al. "Working memory and intelligence: The same or different constructs?" *Psychological Bulletin* 131, 1 (2005): 30–60.

Ambady, N, and R Rosenthal. "Half a minute: Predicting teacher evaluations from thin slices of nonverbal behavior and physical attractiveness." *Journal of Personality and Social Psychology* 64, 3 (1993): 431–441.

Amir, O, et al. "Ha Ha! Versus Aha! A direct comparison of humor to nonhumorous insight for determining the neural correlates of mirth." *Cerebral Cortex* 25, 5 (2013): 1405–1413.

Anderson, ML. *After Phrenology: Neural Reuse and the Interactive Brain*. Cambridge, MA: MIT Press, 2014.

Anguera, JA, et al. "Video game training enhances cognitive control in older adults." *Nature* 501, 7465 (2013): 97–101.

Antoniou, M, et al. "Foreign language training as cognitive therapy for age-related cognitive decline: A hypothesis for future research." *Neuroscience & Biobehavioral Reviews* 37, 10 (2013): 2689–2698.

Arsalidou, M, et al. "A balancing act of the brain: Activations and deactivations driven by cognitive load." *Brain and Behavior* 3, 3 (2013): 273–285.

Bailey, SK, and VK Sims. "Self-reported craft expertise predicts maintenance of spatial ability in old age." *Cognitive Processing* 15, 2 (2014): 227–231.

Bavelier, D. "Your brain on video games." TED Talks, November 19, 2012. https://www.youtube.com/watch?v=FktsFcooIG8.

Bavelier, D, et al. "Brain plasticity through the life span: Learning to learn and action video games." *Annual Review of Neuroscience* 35 (2012): 391–416.

Bavelier, D, et al. "Removing brakes on adult brain plasticity: From molecular to behavioral interventions." *Journal of Neuroscience* 30, 45 (2010): 14964–14971.

Bavishi, A, et al. "A chapter a day: Association of book reading with longevity." *Social Science & Medicine* 164 (2016): 44–48.

Beaty, RE, et al. "Creativity and the default network: A functional connectivity analysis of the creative brain at rest." *Neuropsychologia* 64 (2014): 92–98.

Bellos, A. "Abacus adds up to number joy in Japan." *Guardian*, October 25, 2012. http://www.theguardian.com/science/alexs-adventures-in-numberland/2012/oct/25/abacus-number-joy-japan.

———. "World's fastest number game wows spectators and scientists." *Guardian*, October 29, 2012. http://www.theguardian.com/science/alexs-adventures-in-numberland/2012/oct/29/mathematics.

Benedetti, F, et al. "The biochemical and neuroendocrine bases of the hyperalgesic nocebo effect." *Journal of Neuroscience* 26, 46 (2006): 12014–12022.

Bennett, DA, et al. "The effect of social networks on the relation between Alzheimer's disease pathology and level of cognitive function in old people: A longitudinal cohort study." *Lancet Neurology* 5, 5 (2006): 406–412.

Biggs, J, et al. "The revised two-factor Study Process Questionnaire: R-SPQ-2F." *British Journal of Educational Psychology* 71 (2001): 133–149.

Bloise, SM, and MK Johnson. "Memory for emotional and neutral information: Gender and individual differences in emotional sensitivity." *Memory* 15, 2 (2007): 192–204.

Brewer, JA, et al. "Meditation experience is associated with differences in default mode network activity and connectivity." *PNAS* 108, 50 (2011): 20254–20259.

Buckner, R, et al. "The brain's default network." *Annals of the New York Academy of Sciences* 1124 (2008): 1–38.

Buhle, JT, et al. "Cognitive reappraisal of emotion: A meta-analysis of human neuroimaging studies." *Cerebral Cortex* 24, 11 (2014): 2981–2990.

Burton, R. *On Being Certain*. New York: St. Martin's Griffin, 2008.

Caceres, Z. "The Michael Polanyi College: Is this the future of higher education?" Virgin Disruptors, September 17, 2015. http://www.virgin.com/disruptors/the-michael-polanyi-college-is-this-the-future-of-higher-education.

Chan, YC, and JP Lavallee. "Temporo-parietal and fronto-parietal lobe contributions to theory of mind and executive control: An fMRI study of verbal jokes." *Frontiers in Psychology* 6 (2015): 1285. doi:10.3389/fpsyg.2015.01285.

Channel NewsAsia. "Committee to review Singapore's economic strategies revealed." December 21, 2015. http://www.channelnewsasia.com/news/business/singapore/committee-to-review/2365838.html.

Choi, H-H, et al. "Effects of the physical environment on cognitive load and learning: Towards a new model of cognitive load." *Educational Psychology Review* 26, 2 (2014): 225–244.

Chou, PT-M. "Attention drainage effect: How background music effects concentration in Taiwanese college students." *Journal of the Scholarship of Teaching and Learning* 10, 1 (2010): 36–46.

Clance, PR, and SA Imes. "The imposter phenomenon in high achieving women: Dynamics and therapeutic intervention." *Psychotherapy: Theory, Research & Practice* 15, 3 (1978): 241.

Cognitive Science Online. "A chat with computational neuroscientist Terrence Sejnowski." 2008. http://cogsci-online.ucsd.edu/6/6-3.pdf.

Conway, AR, et al. "Working memory capacity and its relation to general intelligence." *Trends in Cognitive Sciences* 7, 12 (2003): 547–552.

Cooke, S, and T Bliss. "The genetic enhancement of memory." *Cellular and Molecular Life Sciences* 60, 1 (2003): 1–5.

Cotman, CW, et al. "Exercise builds brain health: Key roles of growth factor cascades and inflammation." *Trends in Neurosciences* 30, 9 (2007): 464–472.

Cover, K. *An Introduction to Bridge and Target Technique*. Norfolk: The Syn Alia Animal Training Systems, 1993.

Crick, F. *What Mad Pursuit*. New York: Basic Books, 2008.

Crum, AJ, et al. "Mind over milkshakes: Mindsets, not just nutrients, determine ghrelin response." *Health Psychology* 30, 4 (2011): 424–429.

Davies, G, et al. "Genome-wide association studies establish that human intelligence is highly heritable and polygenic." *Molecular Psychiatry* 16, 10 (2011): 996–1005.

Davis, N. "What makes you so smart, computational neuroscientist?" *Pacific Standard*, August 6, 2015. http://www.psmag.com/books-and-culture/what-makes-you-so-smart-computational-neuroscientist.

Deardorff, J. "Exercise may help brain the most." *Waterbury* (CT) *Republican American*, May 31, 2015. http://www.rep-am.com/articles/2015/06/18/lifestyle/health/884526.txt.

de Bie, T. "Troll Hunting." *Drink a Cup of Tea: And Other Useful Advice on Online Community Management*, December 15, 2013. http://www.tanjadebie.com/ComMan/?p=15.

DeCaro, MS, et al. "When higher working memory capacity hinders insight." *Journal of Experimental Psychology: Learning, Memory, and Cognition* 42, 1 (2015): 39–49.

De Luca, M, et al. "fMRI resting state networks define distinct modes of long-distance interactions in the human brain." *NeuroImage* 29, 4 (2006): 1359–1367.

Deming, WE. *Out of the Crisis*. Cambridge: MIT Press, 1986.

Derntl, B, et al. "Multidimensional assessment of empathic abilities: Neural correlates and gender differences." *Psychoneuroendocrinology* 35, 1 (2010): 67–82.

De Vriendt, P, et al. "The process of decline in advanced activities of daily living: A qualitative explorative study in mild cognitive impairment." *International Psychogeriatrics* 24, 06 (2012): 974–986.

Di, X, and BB Biswal. "Modulatory interactions between the default mode network and task positive networks in resting-state." *PeerJ* 2 (2014): e367.

Dienstmann, G. "Types of meditation: An overview of 23 meditation techniques." *Live and Dare: Master Your Mind, Master Your Life*, 2015. http://liveanddare.com/types-of-meditation/.

DiMillo, I. "Spirit of Agilent." *InfoSpark (The Agilent Technologies Newsletter)*, January 2003.

Dishion, TJ, et al. "When interventions harm: Peer groups and problem behavior." *American Psychologist* 54, 9 (1999): 755–764.

Doherty-Sneddon, G, and FG Phelps. "Gaze aversion: A response to cognitive or social difficulty?" *Memory & Cognition* 33, 4 (2005): 727–733.

Duarte, N. *HBR Guide to Persuasive Presentations*. Cambridge, MA: Harvard Business Review Press, 2012.

Duckworth, A. *Grit*. New York: Scribner, 2016.

Dweck, C. *Mindset*. New York: Random House, 2006.

Dye, MW, et al. "The development of attention skills in action video game players." *Neuropsychologia* 47, 8 (2009): 1780–1789.

———. "Increasing speed of processing with action video games." *Current Directions in Psychological Science* 18, 6 (2009): 321–326.

Einöther, SJ, and T Giesbrecht. "Caffeine as an attention enhancer: Reviewing existing assumptions." *Psychopharmacology* 225, 2 (2013): 251–274.

Eisenberger, R. "Learned industriousness." *Psychological Review* 99, 2 (1992): 248.

Ellis, AP, et al. "Team learning: Collectively connecting the dots." *Journal of Applied Psychology* 88, 5 (2003): 821.

Ericsson, KA, and R Pool. *Peak*. Boston: Eamon Dolan/Houghton Mifflin Harcourt, 2016.

Felder, RM, and R Brent. *Teaching and Learning STEM: A Practical Guide*. San Francisco: Jossey-Bass, 2016.

Fendler, L. "The magic of psychology in teacher education." *Journal of Philosophy of Education* 46, 3 (2012): 332–351.

Finn, ES, et al. "Disruption of functional networks in dyslexia: A whole-brain, data-driven analysis of connectivity." *Biological Psychiatry* 76, 5 (2014): 397–404.

Fox, M, et al. "The human brain is intrinsically organized into dynamic, anticorrelated functional networks." *PNAS* 102 (2005): 9673–9678.

Frank, MC, and D Barner. "Representing exact number visually using mental abacus." *Journal of Experimental Psychology: General* 141, 1 (2012): 134–149.

Freeman, S, et al. "Active learning increases student performance in science, engineering, and mathematics." *PNAS* 111, 23 (2014): 8410–8415.

Friedman, TL. "How to get a job at Google." *New York Times*, February 22, 2014. http://www.nytimes.com/2014/02/23/opinion/sunday/friedman-how-to-get-a-job-at-google.html?_r=0.

Garrison, KA, et al. "Meditation leads to reduced default mode network activity beyond an active task." *Cognitive, Affective, & Behavioral Neuroscience* 15, 3 (2015): 712–720.

Gazzaley, A. "Harnessing brain plasticity: The future of neurotherapeutics." GTC Keynote Presentation, March 27, 2014. http://on-demand.gputechconf.com/gtc/2014/video/s4780-adam-gazzaley-keynote.mp4.

Giammanco, M, et al. "Testosterone and aggressiveness." *Medical Science Monitor* 11, 4 (2005): RA136–RA145.

Golomb, BA, and MA Evans. "Statin adverse effects." *American Journal of Cardiovascular Drugs* 8, 6 (2008): 373–418.

Goyal, M, et al. "Meditation programs for psychological stress and well-being: A systematic review and meta-analysis." *JAMA Internal Medicine* 174, 3 (2014): 357–368.

Green, CS, and D Bavelier. "Action video game training for cognitive enhancement." *Current Opinion in Behavioral Sciences* 4 (2015): 103–108.

Grossman, P, et al. "Mindfulness-based stress reduction and health benefits: A meta-analysis." *Journal of Psychosomatic Research* 57, 1 (2004): 35–43.

Gruber, H. "On the relation between 'aha experiences' and the construction of ideas." *History of Science* 19 (1981): 41–59.

Guida, A, et al. "Functional cerebral reorganization: A signature of expertise? Reexamining Guida, Gobet, Tardieu, and Nicolas' (2012) two-stage framework." *Frontiers in Human Neuroscience* 7 (2013): 590. doi:10.3389/fnhum.2013.00590.

Gwynne, SC. *Empire of the Summer Moon*. New York: Scribner, 2011.

Hackathorn, J, et al. "All kidding aside: Humor increases learning at knowledge and comprehension levels." *Journal of the Scholarship of Teaching and Learning* 11, 4 (2012): 116–123.

Hanft, A. "What's your talent stack?" *Medium*, March 19, 2016. https://medium.com/@ade3/what-s-your-talent-stack-a66a79c5f331#.hd72ywcwj.

Harp, SF, and RE Mayer. "How seductive details do their damage: A theory of cognitive interest in science learning." *Journal of Educational Psychology* 90, 3 (1998): 414.

HarvardX. "HarvardX: Year in Review 2014–2015." 2015. http://harvardx.harvard.edu/files/harvardx/files/110915_hx_yir_low_res.pdf?m=1447339692.

Horovitz, SG, et al. "Decoupling of the brain's default mode network during deep sleep." *PNAS* 106, 27 (2009): 11376–11381.

Howard, CJ, and AO Holcombe. "Unexpected changes in direction of motion attract attention." *Attention, Perception & Psychophysics* 72, 8 (2010): 2087–2095.

Huang, R-H, and Y-N Shih. "Effects of background music on concentration of workers." *Work* 38, 4 (2011): 383–387.

Immordino-Yang, MH, et al. "Rest is not idleness: Implications of the brain's default mode for human development and education." *Perspectives on Psychological Science* 7, 4 (2012): 352–364.

Isaacson, W. "The light-beam rider." *New York Times*, October 30, 2015. http://www.nytimes.com/2015/11/01/opinion/sunday/the-light-beam-rider.html?_r=0.

Jang, JH, et al. "Increased default mode network connectivity associated with meditation." *Neuroscience Letters* 487, 3 (2011): 358–362.

Jansen, T, et al. "Mitochondrial DNA and the origins of the domestic horse." *PNAS* 99, 16 (2002): 10905–10910.

Jaschik, S. "MOOC Mess." *Inside Higher Ed*, February 4, 2013. https://www.inside

highered.com/news/2013/02/04/coursera-forced-call-mooc-amid-complaints-about-course.

Katz, L, and M Rubin. *Keep Your Brain Alive*. New York: Workman, 2014.

Kaufman, SB, and C Gregoire. *Wired to Create*. New York: TarcherPerigee, 2015.

Keller, EF. *A Feeling for the Organism: The Life and Work of Barbara McClintock*, 10th Anniversary Edition. New York: Times Books, 1984.

Kheirbek, MA, et al. "Neurogenesis and generalization: A new approach to stratify and treat anxiety disorders." *Nature Neuroscience* 15 (2012): 1613–1620.

Khoo, A. *Winning the Game of Life*. Singapore: Adam Khoo Learning Technologies Group, 2011.

Kojima, T, et al. "Default mode of brain activity demonstrated by positron emission tomography imaging in awake monkeys: Higher rest-related than working memory-related activity in medial cortical areas." *Journal of Neuroscience* 29, 46 (2009): 14463–14471.

Kühn, S, et al. "The importance of the default mode network in creativity: A structural MRI study." *Journal of Creative Behavior* 48, 2 (2014): 152–163.

Kuhn, T. *The Structure of Scientific Revolutions*. Chicago: University of Chicago Press, 1962 (1970, 2nd ed.).

Li, R, et al. "Enhancing the contrast sensitivity function through action video game training." *Nature Neuroscience* 12, 5 (2009): 549–551.

Lieberman, HR, et al. "Effects of caffeine, sleep loss, and stress on cognitive performance and mood during U.S. Navy SEAL training." *Psychopharmacology* 164, 3 (2002): 250–261.

Lu, H, et al. "Rat brains also have a default mode network." *PNAS* 109, 10 (2012): 3979–3984.

Lv, J, et al. "Holistic atlases of functional networks and interactions reveal reciprocal organizational architecture of cortical function." *IEEE Transactions on Biomedical Engineering* 62, 4 (2015): 1120–1131.

Lv, K. "The involvement of working memory and inhibition functions in the different phases of insight problem solving." *Memory & Cognition* 43, 5 (2015): 709–722.

Lyons, IM, and SL Beilock. "When math hurts: Math anxiety predicts pain network activation in anticipation of doing math." *PLoS ONE* 7, 10 (2012): e48076.

Mantini, D, et al. "Default mode of brain function in monkeys." *Journal of Neuroscience* 31, 36 (2011): 12954–12962.

Maren, S, et al. "The contextual brain: Implications for fear conditioning, extinction and psychopathology." *Nature Reviews Neuroscience* 14, 6 (2013): 417–428.

Markoff, J. "The most popular online course teaches you to learn." *New York Times*, December 29, 2015. http://bits.blogs.nytimes.com/2015/12/29/the-most-popular-online-course-teaches-you-to-learn/.

Marshall, BJ, and JR Warren. "Barry J. Marshall: Biographical." Nobelprize.org, 2005. http://www.nobelprize.org/nobel_prizes/medicine/laureates/2005/marshall-bio.html.

Martin, C. "It's never too late to learn to code." May 7, 2015. https://medium.com/@chasrmartin/it-s-never-too-late-to-learn-to-code-936f7db43dd1.

Martin, D. "Joan McCord, who evaluated anticrime efforts, dies at 73." *New York Times*, March 1, 2004. http://www.nytimes.com/2004/03/01/nyregion/joan-mccord-who-evaluated-anticrime-efforts-dies-at-73.html.

Mazur, A, and A Booth. "Testosterone and dominance in men." *Behavioral and Brain Sciences* 21, 3 (1998): 353–363.

McCord, J. "Consideration of some effects of a counseling program." *New Directions in the Rehabilitation of Criminal Offenders* (1981): 394–405.

———. "Learning how to learn and its sequelae." In *Lessons of Criminology*, edited by Geis, G, and M Dodge, 95–108. Cincinnati: Anderson Publishing, 2002.

———. "A thirty-year follow-up of treatment effects." *American Psychologist* 33, 3 (1978): 284–289.

Mehta, R, et al. "Is noise always bad? Exploring the effects of ambient noise on creative cognition." *Journal of Consumer Research* 39, 4 (2012): 784–799.

Melby-Lervåg, M, and C Hulme. "Is working memory training effective? A meta-analytic review." *Developmental Psychology* 49, 2 (2013): 270–291.

Menie, MA, et al. "By their words ye shall know them: Evidence of genetic selection against general intelligence and concurrent environmental enrichment in vocabulary usage since the mid-19th century." *Frontiers in Psychology* 6 (2015): 361. doi:10.3389/fpsyg.2015.00361.

Merzenich, M. *Soft-Wired*. 2nd ed. San Francisco: Parnassus Publishing, 2013.

Mims, C. "Why coding is your child's key to unlocking the future." *Wall Street Journal*, April 26, 2015. http://www.wsj.com/articles/why-coding-is-your-childs-key-to-unlocking-the-future-1430080118.

Mondadori, CR, et al. "Better memory and neural efficiency in young apolipoprotein E ε4 carriers." *Cerebral Cortex* 17, 8 (2007): 1934–1947.

Montagne, B, et al. "Sex differences in the perception of affective facial expressions: Do men really lack emotional sensitivity?" *Cognitive Processing* 6, 2 (2005): 136–141.

Moon, HY, et al. "Running-induced systemic cathepsin B secretion is associated with memory function." *Cell Metabolism* 24 (2016): 1–9. doi:10.1016/j.cmet.2016.05.025.

Mori, F, et al. "The effect of music on the level of mental concentration and its temporal change." In *CSEDU 2014: 6th International Conference on Computer Supported Education*, 34–42. Barcelona, Spain, 2014.

Moussa, M, et al. "Consistency of network modules in resting-state fMRI connectome data." *PLoS ONE* 7, 8 (2012): e44428.

Nakano, T, et al. "Blink-related momentary activation of the default mode network while viewing videos." *PNAS* 110, 2 (2012): 702–706.

Oakley, B. *Evil Genes: Why Rome Fell, Hitler Rose, Enron Failed, and My Sister Stole My Mother's Boyfriend*. Amherst, NY: Prometheus Books, 2007.

———. "How we should be teaching math: Achieving 'conceptual' understanding doesn't mean true mastery. For that, you need practice." *Wall Street Journal*, September 22, 2014. http://www.wsj.com/articles/barbara-oakley-repetitive-work-in-math-thats-good-1411426037.

———. "Why virtual classes can be better than real ones." *Nautilus*, October 29, 2015. http://nautil.us/issue/29/scaling/why-virtual-classes-can-be-better-than-real-ones.

Oakley, B, et al. "Turning student groups into effective teams." *Journal of Student Centered Learning* 2, 1 (2003): 9–34.

Oakley, B, et al. "Improvements in statewide test results as a consequence of using a Japanese-based supplemental mathematics system, Kumon Mathematics, in an inner-urban school district." In *Proceedings of the ASEE Annual Conference*. Portland, Oregon, 2005.

Oakley, B, et al. "Creating a sticky MOOC." *Online Learning Consortium* 20, 1 (2016): 1–12.

Oakley, BA. "Concepts and implications of altruism bias and pathological altruism." *PNAS* 110, suppl. 2 (2013): 10408–10415.

Oakley, BA. *A Mind for Numbers: How to Excel at Math and Science*. New York: Penguin Random House, 2014.

O'Connor, A. "How the hum of a coffee shop can boost creativity." *New York Times*, June 21, 2013. http://well.blogs.nytimes.com/2013/06/21/how-the-hum-of-a-coffee-shop-can-boost-creativity/?ref=health&_r=1&.

Overy, K. "Dyslexia and music." *Annals of the New York Academy of Sciences* 999, 1 (2003): 497–505.

Pachman, M, et al. "Levels of knowledge and deliberate practice." *Journal of Experimental Psychology: Applied* 19, 2 (2013): 108–119.

Patros, CH, et al. "Visuospatial working memory underlies choice-impulsivity in boys with attention-deficit/hyperactivity disorder." *Research in Developmental Disabilities* 38 (2015): 134–144.

Patston, LL, and LJ Tippett. "The effect of background music on cognitive performance in musicians and nonmusicians." *Music Perception: An Interdisciplinary Journal* 29, 2 (2011): 173–183.

Petrovic, P, et al. "Placebo in emotional processing: Induced expectations of anxiety relief activate a generalized modulatory network." *Neuron* 46, 6 (2005): 957–969.

Pogrund, B. *How Can Man Die Better: Sobukwe and Apartheid*. London: Peter Halban Publishers, 1990.

Powers, E, and HL Witmer. *An Experiment in the Prevention of Delinquency: The Cambridge-Somerville Youth Study*. Montclair, NJ: Patterson Smith, 1972.

Prusiner, SB. *Madness and Memory*. New Haven, CT: Yale University Press, 2014.

Ramón y Cajal, S. *Recollections of My Life*, translated by Craigie, EH. Cambridge, MA: MIT Press, 1989. (Originally published as *Recuerdos de Mi Vida* in Madrid, 1937.)

Rapport, MD, et al. "Hyperactivity in boys with attention-deficit/hyperactivity disorder (ADHD): A ubiquitous core symptom or manifestation of working memory deficits?" *Journal of Abnormal Child Psychology* 37, 4 (2009): 521–534.

Rittle-Johnson, B, et al. "Not a one-way street: Bidirectional relations between procedural and conceptual knowledge of mathematics." *Educational Psychology Review* 27, 4 (2015): 587–597.

Ronson, J. *So You've Been Publicly Shamed*. New York: Riverhead, 2015.

Rossini, JC. "Looming motion and visual attention." *Psychology & Neuroscience* 7, 3 (2014): 425–431.

Sane, J. "Free Code Camp's 1,000+ study groups are now fully autonomous." Free Code Camp, May 20, 2016. https://medium.freecodecamp.com/free-code-camps-1-000-study-groups-are-now-fully-autonomous-d40a3660e292#.8v4dmr7oy.

Sapienza, P, et al. "Gender differences in financial risk aversion and career choices are affected by testosterone." *PNAS* 106, 36 (2009): 15268–15273.

Schafer, SM, et al. "Conditioned placebo analgesia persists when subjects know they are receiving a placebo." *Journal of Pain* 16, 5 (2015): 412–420.

Schedlowski, M, and G Pacheco-López. "The learned immune response: Pavlov and beyond." *Brain, Behavior, and Immunity* 24, 2 (2010): 176–185.

Sedivy, J. "Can a wandering mind make you neurotic?" *Nautilus*, November 15, 2015. http://nautil.us/blog/can-a-wandering-mind-make-you-neurotic.

Shih, Y-N, et al. "Background music: Effects on attention performance." *Work* 42, 4 (2012): 573–578.

Shin, L. "7 Steps to Developing Career Capital and Achieving Success." *Forbes*, May 22, 2013. http://www.forbes.com/sites/laurashin/2013/05/22/7-steps-to-developing-career-capital-and-achieving-success/#256f16d32d3d.

Simonton, DK. *Creativity in Science: Chance, Logic, Genius, and Zeitgeist*. Cambridge, UK: Cambridge University Press, 2004.

Sinanaj, I, et al. "Neural underpinnings of background acoustic noise in normal aging and mild cognitive impairment." *Neuroscience* 310 (2015): 410–421.

Skarratt, PA, et al. "Looming motion primes the visuomotor system." *Journal of Experimental Psychology: Human Perception and Performance* 40, 2 (2014): 566–579.

Sklar, AY, et al. "Reading and doing arithmetic nonconsciously." *PNAS* 109, 48 (2012): 19614–19619.

Smith, GE, et al. "A Cognitive Training Program Based on Principles of Brain Plasticity: Results from the Improvement in Memory with Plasticity-based Adaptive Cognitive Training (IMPACT) Study." *Journal of the American Geriatrics Society* 57, 4 (2009): 594–603.

Snigdha, S, et al. "Exercise enhances memory consolidation in the aging brain." *Frontiers in Aging Neuroscience* 6 (2014): 3–14.

Song, KB. *Learning for Life*. Singapore: Singapore Workforce Development Agency, 2014.

Spain, SL, et al. "A genome-wide analysis of putative functional and exonic variation associated with extremely high intelligence." *Molecular Psychiatry* 21 (2015): 1145–1151. doi:10.1038/mp.2015.108.

Spalding, KL, et al. "Dynamics of hippocampal neurogenesis in adult humans." *Cell* 153, 6 (2013): 1219–1227.

Specter, M. "Rethinking the brain: How the songs of canaries upset a fundamental principle of science." *New Yorker*, July 23, 2001, http://www.michaelspecter.com/wp-content/uploads/brain.pdf.

Stoet, G, and DC Geary. "Sex differences in academic achievement are not related to political, economic, or social equality." *Intelligence* 48 (2015): 137–151.

Sweller, J, et al. *Cognitive Load Theory: Explorations in the Learning Sciences, Instructional Systems and Performance Technologies*. New York: Springer, 2011.

Takeuchi, H, et al. "The association between resting functional connectivity and creativity." *Cerebral Cortex* 22, 12 (2012): 2921–2929.

———. "Failing to deactivate: The association between brain activity during a working memory task and creativity." *NeuroImage* 55, 2 (2011): 681–687.

———. "Working memory training improves emotional states of healthy individuals." *Frontiers in Systems Neuroscience* 8 (2014): 200.

Tambini, A, et al. "Enhanced brain correlations during rest are related to memory for recent experiences." *Neuron* 65, 2 (2010): 280–290.

Teasdale, TW, and DR Owen. "Secular declines in cognitive test scores: A reversal of the Flynn effect." *Intelligence* 36, 2 (2008): 121–126.

Thompson, WF, et al. "Fast and loud background music disrupts reading comprehension." *Psychology of Music* 40, 6 (2012): 700–708.

Tough, P. *How Children Succeed*. Boston: Houghton Mifflin Harcourt, 2012.

Trahan, L, et al. "The Flynn effect: A meta-analysis." *Psychological Bulletin* 140, 5 (2014): 1332–1360.

Tschang, C-C, et al. "50 startups, five days, one bootcamp to change the world." MIT News, August 29, 2014. https://news.mit.edu/2014/50-startups-five-days-one-bootcamp-change-world-0829.

Tupy, ML. "Singapore: The power of economic freedom," Cato Institute, November 24, 2015. http://www.cato.org/blog/singapore-power-economic-freedom.

Vanny, P, and J Moon. "Physiological and psychological effects of testosterone on sport performance: A critical review of literature." *Sport Journal*, June 29, 2015. http://thesportjournal.org/article/physiological-and-psychological-effects-of-testosterone-on-sport-performance-a-critical-review-of-literature/.

Venkatraman, A. "Lack of coding skills may lead to skills shortage in Europe." *Computer Weekly*, July 30, 2014. http://www.computerweekly.com/news/2240225794/Lack-of-coding-skills-may-lead-to-severe-shortage-of-ICT-pros-in-Europe-by-2020-warns-EC.

Vidoni, ED, et al. "Dose-response of aerobic exercise on cognition: A community-based, pilot randomized controlled trial." *PloS One* 10, 7 (2015): e0131647.

Vilà, C, et al. "Widespread origins of domestic horse lineages." *Science* 291, 5503 (2001): 474–477.

Vredeveldt, A, et al. "Eye closure helps memory by reducing cognitive load and enhancing visualisation." *Memory & Cognition* 39, 7 (2011): 1253–1263.

Wager, TD, and LY Atlas. "The neuroscience of placebo effects: Connecting context, learning and health." *Nature Reviews Neuroscience* 16, 7 (2015): 403–418.

Waitzkin, J. *The Art of Learning*. New York: Free Press, 2008.

Wammes, JD, et al. "The drawing effect: Evidence for reliable and robust memory benefits in free recall." *Quarterly Journal of Experimental Psychology* 69, 9 (2016): 1752–1776.

Watanabe, M. "Training math athletes in Japanese jukus." *Juku*, October 21, 2015. http://jukuyobiko.blogspot.jp/2015/10/training-math-athletes-in-japanese-jukus.html.

White, HA, and P Shah. "Uninhibited imaginations: Creativity in adults with attention-deficit/hyperactivity disorder." *Personality and Individual Differences* 40, 6 (2006): 1121–1131.

White, KG, et al. "A note on the chronometric analysis of cognitive ability: Antarctic effects." *New Zealand Journal of Psychology* 12 (1983): 36–40.

Whitehouse, AJ, et al. "Sex-specific associations between umbilical cord blood testosterone levels and language delay in early childhood." *Journal of Child Psychology and Psychiatry* 53, 7 (2012): 726–734.

Wilson, T. *Redirect*. New York: Little, Brown and Company, 2011.

Yang, G, et al. "Sleep promotes branch-specific formation of dendritic spines after learning." *Science* 344, 6188 (2014): 1173–1178.

Zatorre, RJ, et al. "Plasticity in gray and white: Neuroimaging changes in brain structure during learning." *Nature Neuroscience* 15, 4 (2012): 528–536.

Zhang, J., and X Fu. "Background music matters: Why strategy video game increased cognitive control." *Journal of Biomusical Engineering* 3, 105 (2014): doi:10.4172/2090-2719.1000105.

Zhao, Y. *Who's Afraid of the Big Bad Dragon*. San Francisco: Jossey-Bass, 2014.

Zhou, DF, et al. "Prevalence of dementia in rural China: Impact of age, gender and education." *Acta Neurologica Scandinavica* 114, 4 (2006): 273–280.

Zittrain, J. "Are trolls just playing a different game than the rest of us?" *Big Think*, April 3, 2015. http://bigthink.com/videos/dont-feed-the-trolls.

Zull, JE. *The Art of Changing the Brain*. Sterling, VA: Stylus Publishing, 2002.

Notes

Chapter 1: Transformed

1. Dweck, 2006.

Chapter 2: Learning Isn't Just Studying

1. Deardorff, 2015.
2. Deardorff, 2015. See also Cotman, et al., 2007; Moon, et al., 2016.
3. Snigdha, et al., 2014.
4. Vidoni, et al., 2015.

Chapter 3: Changing Cultures

1. Gwynne, 2011.
2. Mims, 2015; Venkatraman, 2014.
3. Ericsson and Pool, 2016.
4. Yang, 2014.
5. Oakley, "How we should be teaching math," 2014; Rittle-Johnson, et al., 2015.
6. Nobel Prize winners' biographies and autobiographies are often replete with stories of resistance to new ideas and approaches. See, for example, Ramón y Cajal, 1989; Keller, 1984; Prusiner, 2014; Marshall and Warren, 2005. For a great review of antagonism by scientific leaders toward the idea of neurogenesis, see Specter, 2001.
7. Kuhn, 1962 (1970, 2nd ed.), 144.

Chapter 4: Your "Useless" Past Can Be an Advantage

1. Tanja's games are Tazlure.nl (fantasy) and a historical seventeenth-century court-of-England game. Tanja asked me not to post the URL of the latter here because enrollment is limited, but the sophisticated game draws you in from the opening page on.
2. de Bie, 2013. For an academically oriented take that gives a sense of how little academics know about trolls, see Zittrain, 2015. An excellent discussion of the dark world of troll-dom can be found in Ronson, 2015.
3. Stoet and Geary, 2015; Whitehouse, et al., 2012.
4. These images are meant to give a figurative sense of key ideas contained in Stoet and Geary, 2015, and Whitehouse, et al., 2012.
5. Vanny and Moon, 2015.

Chapter 5: Rewriting the Rules

1. For five years, I volunteered in a number of elementary schools in the inner-urban school district of Pontiac, Michigan, and experienced firsthand the conditions endured by students in a typical disadvantaged school district (Oakley, et al., 2005).
2. I met Stephanie Caceres, Zach's mother, for tea in Linthicum Heights, Maryland, on May 12, 2016.
3. McCord, 2002.
4. Dishion, et al., 1999, 760.
5. Powers and Witmer, 1972. See in particular Chapter 29.
6. McCord, 1981; McCord, 1978.
7. Ibid.
8. Rittle-Johnson, et al., 2015.
9. Guida, et al., 2013.
10. Pachman, et al., 2013.
11. Caceres, 2015.
12. McCord, 1978.
13. E-mail correspondence with Geoff Sayre-McCord, June 2016.
14. Martin, 2004.
15. Wilson, 2011.
16. Duckworth, 2016.
17. Eisenberger, 1992.
18. Oakley, 2013; Wilson, 2011.
19. Fendler, 2012.
20. Tough, 2012.

Chapter 6: Singapore

1. Song, 2014.
2. Trading Economics, "Singapore Unemployment Rate, 1986–2016." http://www.tradingeconomics.com/singapore/unemployment-rate.

3. Tupy, 2015.
4. National Center for Education Statistics, "Mathematics Literacy: Average Scores." https://nces.ed.gov/surveys/pisa/pisa2012/pisa2012highlights_3a.asp, citing Organization for Economic Cooperation and Development (OECD), Program for International Student Assessment (PISA), 2012.
5. Hanft, 2016.
6. Shin, 2013.
7. Professor Yong Zhao, who holds the first presidential chair at the University of Oregon, notes apropos mainland China: "Chinese students are extremely good at well-defined problems. That is, as long as they know what they need to do to meet the expectations and have examples to follow, they do well. But in less defined situations, without routines and formulas to fall back on, they have great difficulty. In other words, they are good at solving existing problems in predictable ways, but not at coming up with radical new solutions or inventing new problems to solve" (Zhao, 2014, 133–134). Zhao also has a lengthy discussion of the problems with PISA testing in Chapter 8, "The Naked Emperor: Chinese Lessons for What Not to Do."
8. Channel NewsAsia, 2015.

Chapter 7: Leveling the Educational Playing Field

1. Wammes, et al., 2016.
2. Sklar, et al., 2012.
3. Bellos, "World's fastest number game," 2012.
4. Watanabe, 2015. See also "Begin Japanology—Abacus," https://www.youtube.com/watch?v=zMhcr—d6bw.
5. Bellos, "Abacus adds up," 2012.
6. Frank and Barner, 2012.
7. Guida, et al., 2013.
8. Ericsson and Pool, 2016.
9. Arsalidou, et al., 2013; Sweller, et al., 2011.
10. Guida, et al., 2013.
11. See also Chapter 5, Khoo, 2011.
12. Commencement address delivered by Steve Jobs at Stanford University, June 12, 2005, http://news.stanford.edu/2005/06/14/jobs-061505/.
13. Buhle, et al., 2014.
14. Focused mode is often referred to as "task positive" in the literature. See Di and Biswal, 2014; Fox, et al., 2005.
15. The best-studied of the many neural resting states, of course, is the default mode network. Moussa, et al., 2012.
16. Beaty, et al., 2014.
17. Nakano, et al., 2012.
18. Waitzkin, 2008, 159.
19. Tambini, et al., 2010; Immordino-Yang, et al., 2012.
20. Brewer, et al., 2011; Garrison, et al., 2015.
21. Immordino-Yang, et al., 2012.

22. For a sampling of the range of recent findings, see Garrison, et al., 2015, as opposed to Jang, et al., 2011. Clearly this is a complex area, with varying effects. A nice overview of meditation techniques, delineating which types are open monitoring and which are focused attention, can be found at Dienstmann, 2015. See also Kaufman and Gregoire, 2015, 110–120; Goyal, et al., 2014; Grossman, et al., 2004.
23. Sedivy, 2015.
24. Ackerman, et al., 2005; Conway, et al., 2003.
25. DeCaro, et al., 2015.
26. Lv, 2015; Takeuchi, et al., 2012; White and Shah, 2006.
27. Patros, et al., 2015; Rapport, et al., 2009.
28. Simonton, 2004.
29. See Ellis, et al., 2003, who note: "Agreeable team members, who by definition are compliant and deferent, may more readily accept the opinion of their team members uncritically in order to avoid argument."
30. Melby-Lervåg and Hulme, 2013.
31. Smith, et al., 2015—there's about a 4 percent increase which, as related research shows, appears to be lasting. A list of up-to-date research publications on BrainHQ's memory-related programs is maintained here: http://www.brainhq.com/world-class-science/published-research/memory.
32. Takeuchi, et al., 2014.
33. One American with long experience in Asia told me, "It's almost like people don't even want to go to college unless it's at a top school. In America people typically apply to backup schools and just go there if they don't get into their dream school. In Asia it's not uncommon for students to take one or more years off and prepare to take the exams again if they didn't get in. Entrance decisions are determined more by a battery of entrance exams they take than the typical U.S. college application where an SAT score is just one part of your overall app. There are full- and part-time test schools specifically for these kinds of students. There's even a word in Korean for these students—*jaesuseng*. You often see reality show type things on TV about kids who are in their fourth or fifth year of retrying the exams. Will they finally get in? Will they give up? That kind of thing. It's a real cultural phenomenon. People certainly do still go to second-tier schools, and I think some people genuinely are embarrassed about it. But whenever you see the topic discussed in the media or academia it's almost like these second tier schools don't even exist. They are never talked about. The issue is always framed in terms of the top schools being too competitive and driving unhealthy primary and secondary school culture. Just simply having a college education does not appear to be respected in its own right. For lack of a better way to put it, there needs to be greater social acceptance of being average. Academic credentials have far reaching effects in the workplace too. In America I'm at a point where if I were applying to a job I'd list my work experience first and academic credentials second. But in Asia I'd put my Stanford degree front and center."

Chapter 8: Avoiding Career Ruts and Dead Ends

1. The discussions of Terrence Sejnowski's past grow from an extended interview with Terry and his wife, Beatrice Golomb, on July 26, 2015, in La Jolla, California.
2. Davis, 2015.
3. Interview with Alan Gelperin, March 5, 2016, Princeton, New Jersey.
4. *Cognitive Science Online*, 2008.
5. Ibid.
6. Golomb and Evans, 2008.
7. Maren, et al., 2013.
8. Benedetti, et al., 2006.
9. Wager and Atlas, 2015.
10. Schafer, et al., 2015. Petrovic, et al., 2005 note, "placebo has been shown to be crucially dependent on learning effects."
11. Crum, et al., 2011.
12. Schedlowski and Pacheco-López, 2010.
13. Petrovic, et al., 2005.
14. Crick, 2008, 6.
15. Bavelier, et al., 2010; Dye, et al., "The development of attention skills in action video game players," 2009; Dye, et al., "Increasing speed of processing with action video games," 2009; Green and Bavelier, 2015; Li, et al., 2009. In support of this section's discussion, see also many earlier studies by these and related authors.
16. Bavelier, et al., 2012.
17. Howard and Holcombe, 2010; Lv, et al., 2015; Rossini, 2014; Skarratt, et al., 2014.
18. Bavelier, 2012.
19. Anguera, et al., 2013.
20. Gazzaley, 2014.
21. Beta waves are more often associated with focused attention. In this regard, Dr. Gazzaley notes: "The theta waves that we look at are locked in time to an event, in the case of *Neuroracer* to the appearance of a sign while driving. And it bursts after the appearance of the sign. This type of theta is associated with engagement of attention. It differs from theta that is engaged more tonically" (E-mail to the author, June 2, 2016).
22. Gazzaley, 2014.
23. Merzenich, 2013, 197.
24. Posit Science Corporation keeps a list of the scientific research that supports claims of the efficacy of their therapies: http://www.brainhq.com/world-class-science/published-research.
25. Spalding, et al., 2013.
26. Ibid.
27. Kheirbek, et al., 2012.
28. Ibid.
29. Katz and Rubin, 2014.

30. Antoniou, et al., 2013.
31. Pogrund, 1990, 303.
32. See White, et al., 1983. My husband, Philip, who spent a year with an eight-man crew isolated at remote Siple Station in Ellsworth Land in Antarctica, observed this personally.
33. De Vriendt, et al., 2012.
34. Bailey and Sims, 2014.
35. Bavishi, et al., 2016.
36. Zhou, et al., 2006.
37. Bennett, et al., 2006.
38. Davis, 2015.

Chapter 9: Derailed Dreams Lead to New Dreams

1. Princess's story is based on her recollections, as told to me via e-mail in the May to July 2016 time frame.
2. Clance and Imes, 1978.
3. Bloise and Johnson, 2007; Derntl, et al., 2010; Montagne, et al., 2005.
4. Sapienza, et al., 2009; Mazur and Booth, 1998; Giammanco, et al., 2005.
5. Ramón y Cajal, 1937 (reprint 1989).
6. Burton, 2008.

Chapter 10: Turning a Midlife Crisis into a Midlife Opportunity

1. DiMillo, 2003. These descriptions of Arnim Rodeck and his experiences came from e-mail interviews and essays Arnim provided to me during May and June 2016.
2. See Ericsson and Pool, 2016, 222–225, for a discussion of "anti-prodigies." Ericsson notes that in cases where someone appears to have no talent, it is usually some early authority figure who convinces them of it. True tone-deafness, he observes, is vanishingly rare. On the other hand, it is clear that some individuals have an underlying neural structure that can make some tasks difficult for them to learn. For example, Finn, et al., 2014, observe, "Compared to [non-impaired] readers, [dyslexic] readers showed divergent connectivity within the visual pathway and between visual association areas and prefrontal attention areas. . . ." Could it be that Arnim's dyslexia is relevant with respect to his challenges with music? Indeed, researchers have found that those with dyslexia often have timing deficits that affect their musical abilities (Overy, 2003). It's interesting to contrast Arnim's reaction with relation to the authority figure in music (acceptance of his seeming inability with music and finding another way around the obstacle), versus his reaction to the authority figure in math (head-on tackling of math to improve his abilities). My personal belief is that whatever the provenance of underlying neural structures, even if they might make learning more difficult, those differing structures can help make for a different, deeper, more creative understanding if the learner finds a path through the mental obstacles.
3. Mehta, et al., 2012; O'Connor, 2013.

4. Einöther and Giesbrecht, 2013; Lieberman, et al., 2002. When we are in thinking mode (which is hopefully a fair bit of the time), all brain wave frequencies are present, but usually only one band of frequencies is dominant depending on our state of consciousness. Interestingly, ADHD is affiliated with more activity in the longer wavelength bands like alpha and theta, whereas focused attention is more affiliated with the higher gamma frequencies.
5. Choi, et al., 2014; Doherty-Sneddon and Phelps, 2005.
6. Vredeveldt, et al., 2011.
7. Cooke and Bliss, 2003; Davies, et al., 2011; Spain, et al., 2015; Mondadori, et al., 2007. Exceptional memories can be a powerful tool that can help boost people to top leadership roles. See Oakley, 2007, 310–314.
8. An excellent discussion of the yin-yang operation of the two different modes (that is, one mode is generally active while the other is at rest) is in Sinanaj, et al., 2015.
9. De Luca, et al., 2006.
10. Kühn, et al., 2014; Takeuchi, et al., 2011.
11. Gruber, 1981. As noted in Horovitz, et al., 2009, default mode network connectivity persists during light sleep: "[T]his persistence could be expected because self-reflective thoughts do not abruptly cease but rather decrease gradually as a person falls asleep, to the point of being virtually absent during the deepest stages of sleep."
12. Buckner, et al., 2008; O'Connor, 2013.
13. Sinanaj, et al., 2015.
14. Patston and Tippett, 2011; Thompson, et al., 2012; Chou, 2010. Video game designers have capitalized on the fact that a bit of background music appears to help video gamers concentrate better when they're strategizing about what to do next in complex situations ("proactive control"), while that same music can detract from simple reactions ("reactive control") (Zhang and Fu, 2014).
15. Shih, et al., 2012.
16. Huang and Shih, 2011; Mori, et al., 2014.
17. Arnim's website is at www.shamawood.com. Warning, though—his work is in high demand!

Chapter 11: The Value of MOOCs and Online Learning

1. The individuals described and quoted in this chapter were interviewed via e-mail in the April through July 2016 time frame.
2. http://davidventuri.com/blog/my-data-science-masters.
3. Check out Brian's progress at http://www.brianbrookshire.com/online-biology-curriculum/.
4. A pseudonym per "Hans's" request.
5. Target Test Prep, https://gmat.targettestprep.com/.
6. Ronny takes the process improvement approach of "plan-do-check-adjust" for his own learning efforts. "PDCA" was first developed by W. Edwards Deming, the father of modern quality control (Deming, 1986).
7. For a great discussion on the passive nature of television viewing and its impact on learning, see Zull, Chapter 3. Interestingly, despite the strong emphasis on

active learning, and the fact that active learning has been shown to be critically important in the classroom (Freeman, et al., 2014; Oakley, et al., "Turning student groups into effective teams," 2003), there is little by way of neuroimaging research that provides insight into what is going on in active as opposed to passive learning. Indeed, understanding what learning does to the brain is still in its infancy, although it's definitely a hot topic of research (Zatorre, et al., 2012).

8. Biggs, et al., 2001, provide an interesting attempt to get at whether students are approaching their learning using a "deep" versus "surface" approach.

Chapter 12: MOOC-Making

1. See Markoff, 2015, which notes 1,192,697 registered students for "Learning How to Learn" from August 1, 2014, to December 2015. By contrast, from roughly 2012 to June 2015 HarvardX had had 60-plus open courses/modules, 17-plus on-campus courses, 7-plus SPOCs (small private online courses), with 90 Harvard faculty, and 225 other individuals, which altogether encompassed more than 3 million course registrations (HarvardX, 2015). In other words all 84 of Harvard's MOOC's and other online courses were bringing in around 83,300 students per month over a three-year period, while "Learning How to Learn" was running close behind, bringing in 70,200 per month in the first 17 months from its beginning. Not too shabby for an upstart little course made in a basement!
2. To create the "Learning How to Learn" MOOC, it initially took about four months of half-time work to set up the studio, learn how to edit video, and create the first few disastrous videos that ended up being thrown away. Then it took another three months of full-time scripting, filming, and editing, often fourteen hours a day, along with developing the quiz questions and rubrics.

   I advise using green screen where you can because it gives you lots more ability to add motion. You can pop your "talent" back and forth and from full stand to close-up. This activates any number of layers of attentional mechanisms. (See Oakley, 2015, and especially Oakley, et al., 2016, for background about the neuroscience involved in maintaining attention.)

   Here are a few special tips on what to do—and what not to do—when MOOC-making:

   - Use a high shutter speed, something like 80. This will prevent blurry green from the green screen from showing through your fingers.
   - For the green screen, use four lights instead of three when you set up your studio lighting. This will allow for the higher shutter speed you need to prevent that blurry green from showing through your fingers, as noted above.
   - Set your focus very carefully when using green screen. Do this by using the magnifier to focus in on the wrinkles that will probably be at the side of the eye of your talent. (Unless your talent is, like, two years old.) Whenever you walk away from your camera, recheck the focus—it can drift out quite easily.
   - You will probably have a lavalier microphone that clips to your lapel. Directly where the mike clips to your lapel, make a little loop of the wire and clip to both the loop and the lapel. This will relieve the mechanical tension on the wire. If

you don't do this, you'll get all sorts of rustling sounds that are terribly difficult, if not impossible, to get rid of in postproduction.

- Don't worry that your first video footage will probably show you looking like you're standing in front of a firing squad. That's perfectly normal and will disappear once you start getting ~~bored~~ used to being in front of the camera. (If you weren't nervous for the first few days in front of a camera, I'd wonder what your scores on the Hare Psychopathy Checklist might be.)
- Editing yourself might seem to be a task best left to others. But it's important to learn at least modest proficiency with video editing. You'll be more creative with the possibilities. Editing footage of yourself is particularly helpful if you are nervous in front of the camera. At first you'll be hypercritical and it will seem like an antitherapeutic exercise. But after a while you'll begin to realize, when you start watching television news, for example, that even the pros make the same "mistakes" you criticize yourself for making. Editing yourself helps extinguish excessive concerns about how you look, because after a while, you just get bored with the hypercriticism.
- Watch out that your voice can get higher and squeakier when you're nervous. Since women start out at a higher register to begin with, if they're not careful, they can end up unpleasantly chipmunk-like. Whether you're a man or woman, unless you happen to naturally sound like Johnny Cash, you may wish to practice speaking with a somewhat deeper voice.
- Do not wear a white bra because it will show through your blouse—always wear beige. (Don't worry about this if you don't wear a bra.) And incidentally, pearls can look wonderful, but they also can bump against the microphone and create annoying sounds, so it's best to avoid them.
- I was using a teleprompter, and I'd often want to go back to the beginning of a five-minute script if I flubbed anywhere, even at the end. Do not do this. You're going to end up cutting anyway, to provide for motion. Just go back to the start of the sentence, paragraph, or beginning of the train of thought.

3. Jaschik, 2013.
4. Ambady and Rosenthal, 1993.
5. Some people like working with scripts, and others with teleprompters. Most instructors inevitably hesitate in places when they're speaking extemporaneously. The challenge with scripted approaches, on the other hand, is that it's easy to write something pedantic that's a real snooze to listen to. Worse yet, Some. Professors. Use. Teleprompters. To. Speak. Like. This.
6. Lyons and Beilock, 2012.
7. Amir, et al., 2013.
8. Chan and Lavallee, 2015.
9. Hackathorn, et al., 2012.
10. The faulty nature of the "we've got too much to cover" idea, particularly in STEM teaching, is discussed at length in Felder and Brent, 2016.
11. The research most frequently cited about humor impeding learning is Harp and Mayer, 1998. Interestingly, however, this relates to written materials—not "in

vivo" teaching by a live instructor or a video. Ironically, Richard Mayer, the coauthor of this research, is one of the wittiest public speakers you'd ever hope to find.

12. Rossini, 2014; Skarratt, et al., 2014.
13. Oakley, et al., 2016; Oakley and Sejnowski, 2016; Rossini, 2014; Skarratt, et al., 2014.
14. Anderson, 2014. Metaphor can be terrifically important in learning—see Oakley, 2014, Chapter 11; Oakley, et al., 2016; Oakley and Sejnowski, submitted, along with the embedded references.
15. Keller, 1984.
16. Isaacson, 2015.
17. Sane, 2016.
18. Tschang, et al., 2014.
19. Trahan, et al., 2014. But see Menie, et al., 2015, and Teasdale and Owen, 2008, for evidence of more recent leveling or decline.
20. Duckworth, 2016, 84.
21. Duarte, 2012.

Chapter 13: Mindshift and More

1. Both "Louise's" and "Specs's" names have been changed by mutual agreement—various identifying details of their life together today have also been changed. Descriptions of Specs's early years before he met Louise have been fictionalized, although they are based on descriptions of people and horses I knew growing up with horses as the daughter of a veterinarian who focused on cattle and horses. It's probably worth mentioning that some decades ago friends of mine and I created the popular, perennially bestselling educational board game about horses Herd Your Horses. Maybe it runs in the blood—my mom's father, Clarence C. Pritchard, was a rancher and apparently a well-known local "horse whisperer" near Roswell, New Mexico.

   I have spent time in person together with Louise and Specs at Louise's house and horse facilities, and can attest to the fact that Specs is an amazing horse who is able to do the types of activities described in this chapter.
2. Evidence suggests that animals very similar to modern horses have been around for some three hundred thousand years (Jansen, et al., 2002).
3. Vilà, et al., 2001.
4. Friedman, 2014.
5. Kojima, et al., 2009; Lu, et al., 2012; Mantini, et al., 2011.
6. Cover, 1993.

Index

Also by
Barbara Oakley, PhD

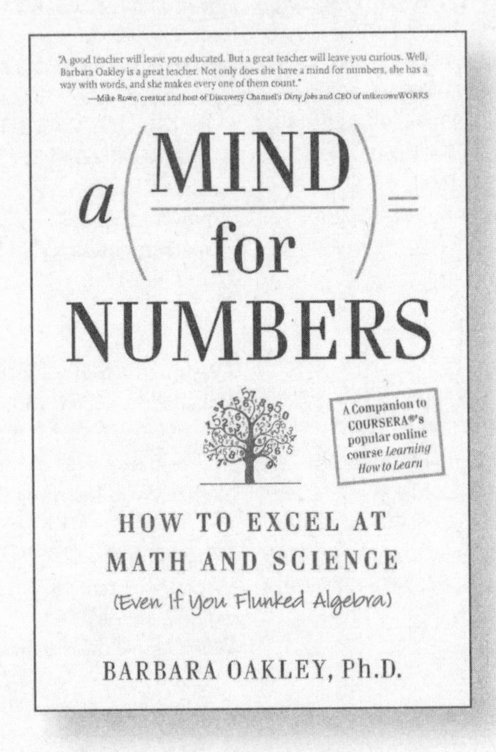

The companion book to COURSERA®'s wildly popular massive open online course "Learning How to Learn"

"A good teacher will leave you educated. But a great teacher will leave you curious. Well, Barbara Oakley is a great teacher. Not only does she have a mind for numbers, she has a way with words, and she makes every one of them count."

—Mike Rowe, creator and host of Discovery Channel's *Dirty Jobs* and CEO of mikeroweWORKS